THE STRUCTURE OF THE ATMOSPHERE

IN CLEAR WEATHER

THE STRUCTURE OF THE ATMOSPHERE
IN CLEAR WEATHER

A STUDY OF SOUNDINGS
WITH PILOT BALLOONS

BY

C. J. P. CAVE, M.A.

Nonne vides etiam diversis nubila ventis
Diversas ire in partis inferna supernis?

LUCRETIUS, v. 646

Cambridge:
at the University Press
1912

CAMBRIDGE
UNIVERSITY PRESS

University Printing House, Cambridge CB2 8BS, United Kingdom

Cambridge University Press is part of the University of Cambridge.

It furthers the University's mission by disseminating knowledge in the pursuit of education, learning and research at the highest international levels of excellence.

www.cambridge.org
Information on this title: www.cambridge.org/9781107457980

© Cambridge University Press 1912

First published 1912
First paperback edition 2014

A catalogue record for this publication is available from the British Library

ISBN 978-1-107-45798-0 Paperback

CONTENTS

LIST OF ILLUSTRATIONS

INTRODUCTION

THE investigation of the wind currents of the air above the surface layers is one of the greatest importance in the study of meteorology; one reason for the slow advance made by this science in the last fifty years is to be found in the fact that until quite recently meteorologists only took note of that part of the atmosphere that was close to the surface of the earth, and beyond some cloud observations and a few isolated records, such as those obtained by Glaisher, nothing was known of the conditions existing in the free air. The recent rise of aviation and its probable extension in the near future make it more than ever necessary to investigate the nature of the currents in the free air above the surface of the earth.

During the last few years the conditions of temperature, humidity, and wind have been investigated by means of kites carrying self-recording instruments to very considerable heights. Free balloons carrying lighter instruments have continued these records to still higher regions, heights of 25 kilometres and more having been reached. The motion of such a balloon if accurately observed gives a record of the wind currents traversed by it in its ascent through the atmosphere. Such records may also be obtained by small balloons that carry no instrument when they are followed by means of a theodolite during their ascent. The following pages give some account of the investigation of the upper air by means of such observations, some of the records having been obtained from balloons carrying instruments and others from small free balloons carrying nothing beyond a stamped label to be posted if the burst balloon should be found after it reaches the earth. An account is given in the first chapter of the general types of structure disclosed by the observations, and figures are given of models prepared to show the sequence of wind velocities and directions met with during the ascents on occasions when the different types of structure were found. An account follows of the methods of observing the balloons and of the theodolites employed for this purpose, together with an account of how the observations are worked up to give the horizontal trajectory of the balloon, and the method of measuring the wind velocity and direction at different heights from the trajectory. In Chapter III will be found a discussion of the accuracy of the methods employed, and a comparison of the trajectory determined by the observations of two theodolites at opposite ends of a base line with that determined by the observations of one theodolite only, and the assumption that the balloon ascends with a uniform or at any rate with a known velocity. Following on this is a discussion of the rate of ascent of rubber balloons which it is of great importance to determine as accurately as possible; in this connection the results of observations and of theoretical considerations by investigators in this country and on the Continent are given. The relation of surface air currents to the configuration of the ground is also touched on; this is a point of great importance for aviators and it is one that should be gone into more fully with balloons that ascend more slowly than those that have been used in these investigations.

A general summary of the results obtained is given in Chapter V in which certain types of structure in the atmosphere are recognised, and the different types are considered in their relation to the wind at the surface, the gradient wind, and the general distribution of pressure and temperature in the region. Five types are described: (a) wind in the upper air steady with no increase in velocity with height; (b) wind in the upper air increasing, sometimes to several times the gradient value, but remaining more or less steady in direction; (c) wind in the upper air decreasing in velocity; (d) reversals or great changes in wind direction in the upper air; (e) wind in the upper air blowing away from centres of low pressure. In the types represented by these five classes the wind in the upper air has been compared with that on the surface. A consideration of the higher ascents has shown that the strongest current is as a rule to be found in the region just below the stratosphere. This rapidly moving current must be associated with a corresponding pressure distribution in that region. Recent researches[1] have tended to show that it is there that changes of pressure originate, and from this point of view the layer just below the stratosphere must be regarded as controlling the conditions throughout the atmosphere beneath. The transference of the supposed seat of action from the surface to the region of nine kilometres suggests that variations in the currents in the layers beneath might with advantage be referred to the conditions prevailing at the time at the nine kilometre level instead of to those at the surface. This method of looking at the results of the ascents was suggested to me by Dr W. N. Shaw when this book was already in type. An examination of the cases which are represented by diagrams at the end of the volume shows that the method would greatly simplify the systematic representation of the atmospheric stratum between the surface and the region in question. Starting with a strong Westerly wind under the stratosphere we find almost without exception that the Westerly wind falls off in the lower levels, and the falling off may proceed continuously to such an extent that the direction

[1] See W. H. Dines, F.R.S., "Statical Changes of Pressure and Temperature in a Column of Air that accompany Changes of Pressure at the bottom," *Quart. Journ. Royal Meteorological Society*, vol. XXXVIII. p. 41; and letters in *Nature*, vol. LXXXVIII. p. 141 by Dr W. N. Shaw, F.R.S., and p. 175 by Mr W. H. Dines.

of motion is reversed at some point in the intermediate layers, so that near the surface an Easterly wind is shown instead of the Westerly one of the upper regions. Even if the intermediate layers themselves provided no variation in the distribution of pressure that would affect the velocity we should expect the strength of the current to be diminished in the lower layers because the density there is greater than in the higher regions, and the velocity corresponding to the pressure gradient transmitted from above would be less in the inverse proportion of the density; on this ground alone the wind velocity near the surface would be reduced to about one-third of the velocity at nine kilometres. But in the cases considered it will be evident that the diminution in the Westerly wind is at a greater rate than the increase of density, and the additional decrease must be due to pressure distribution accruing in the lower layers. Actual reversals are accounted for by representing the additional decrease as a superposed Easterly wind originating from pressure distribution in the lower layers which is sometimes so great as to show a wind at the surface in a reversed direction. The gradual modification of the gradient with increasing depth below the nine kilometre layer could easily be accounted for by a distribution of temperature in the layers underneath such that the air to the North is always colder than the air to the South, that is to say by assuming a distribution of temperature that corresponds with the latitude. The more rapid reversals on special occasions would thus be accounted for by a fall of temperature from South to North greater than the average. It will be seen that from this point of view the reversal of the air current implies no discontinuity in the atmosphere below the nine kilometre level, the change from West to East taking place gradually throughout the whole thickness.

This result is of general application in all the high ascents either as a decrease of the Westerly wind as described, or in some cases as an absence of decrease of an Easterly wind if such should exist at the nine kilometre level, when the decrease of velocity due to the density as the surface is approached is balanced by the increase due to the pressure distribution in the lower layers. This effect of the lower atmosphere in producing an Easterly component of the wind which is stronger the nearer to the surface is quite in accord with the calculations of M. Teisserenc de Bort of an average Westerly circulation in the four kilometre level modified as regards the lower layers by the distribution of temperature.

Similarly regularity is not apparent as regards the winds from North and South, and the recognition of this fact has led, on Dr Shaw's suggestion, to an examination of a number of the ascents by the analysis of the wind at each level into a West-East component and a South-North component. This process has simplified the classification of the ascents in a remarkable manner. It appears that the structure of the atmosphere as disclosed by all the high ascents can be represented as regards the West-East component by the gradual development of an East-West component increasing continuously as the surface is approached, and doubtless due to the temperature distribution in latitude. As regards the South-North component the effect of the lower layers is to alter the velocity by the continuous addition of a component which may be from the North or from the South according to circumstances. The South-North component shows a decrease of intensity as the surface is approached but there is no differentiation between the effect of the lower layers such as that shown by the West-East component. Thus the variation in the Northerly and Southerly winds depends on meteorological conditions which may show effects in opposite directions on different occasions. The effect of the layers beneath the nine kilometre level may be seen in the ascent for Nov. 6th, 1908, when the West-East component at nine kilometres was 11 metres per second; the effect of the superposed East-West component due to the lower layers was to reduce the velocity fairly regularly till at 3·5 kilometres the West-East component was balanced by the East-West component; at lower layers there was a reversal and at one kilometre above sea level the East-West component was 13 metres per second, or perhaps it is clearer to say that the West-East component was −13 metres per second. At the same time the South-North component had increased from −5 to +9 metres per second. On Oct. 1st, 1908, the West-East component decreased from +11 metres per second at 9 kilometres to −1 at 4 kilometres, below which however the decrease was not maintained; the South-North component decreased from 18 to 13 metres per second, the decrease continuing down to the ground level. One more example may be given, Sept. 15th, 1911, an ascent not elsewhere discussed in this book; the West-East component decreased from +32 metres per second at 9 kilometres to −8 at 1 kilometre while the South-North component decreased from +12 to −10 metres per second.

The gradual increases which are here described may be distinguished from the occasional increases of velocity locally at different times at various levels which appear as protruberances on the curve of relation of velocity of the several components with height. With these localised disturbances may probably be grouped the remarkably rapid variations with velocity shown in the lower layers on some of the occasions when the balloon was lost to sight on account of clouds at a comparatively low level. For these disturbances no explanation is offered for the present.

The foregoing considerations which did not suggest themselves till this book was already in type should be born in mind in Chapter IX which deals with the relation of vertical wind distribution to the distribution of pressure at the surface.

The subject of the wind in the stratosphere forms the subject of a separate chapter. It is quite clear that when a balloon enters this region it meets with winds of much smaller intensity than those traversed below this level. With Westerly, Northerly, and Southerly winds the stratosphere wind as far as has been observed remains more or less the same in direction as the winds in the lower strata, though with greatly decreased velocity; but when Easterly winds are found in the layers immediately below the stratosphere the wind in that region exhibits curious fluctuations; the balloon trajectory traces out loops as though spiral motions were met with. Since the observations dealt with in this book were concluded several more balloons have been observed till they were well within the stratosphere, and these observations fully bear out what is herein recorded.

When the balloon enters the stratosphere the West-East component decreases, as also however do the South-North or North-South components; a decrease in the West-East component would be occasioned if the air at this level were colder to the South than to the North, that is if there were a temperature gradient in latitude in the reverse direction to that at the surface; this is probably the case; observations in low latitudes by M. Teisserenc de Bort and Professor A. Lawrence Rotch have shown that at heights above nine or ten kilometres the temperatures in the low latitudes are lower than the temperatures at corresponding heights in higher latitudes.

At the end of the book will be found two tables; the first gives a list of the 200 ascents in order of date with the greatest height to which the observations were carried in each instance, to what class each ascent belongs, and the distance of the point of fall where this is known. The second table commencing on page 84 gives the wind velocity and direction for each ascent for every half kilometre of height, and at the beginning of each ascent will be found the gradient velocity and direction in all cases when the gradient was sufficiently definite for this to be calculated. I have to thank Mr R. Corless and Mr R. G. K. Lempfert, members of the staff of the Meteorological Office, for kindly giving me the necessary information about the gradient wind at the times of the balloon ascents.

After the tables will be found 24 diagrams giving the wind velocity and direction plotted against the height for certain typical or interesting ascents, together with weather maps showing the isobars and the velocity and direction of the wind at the surface, information which was taken from the Weekly Weather Report of the Meteorological Office. Diagrams showing the variation of the wind with height have been prepared for all of the 200 ascents, but it was not found practicable to reproduce more than those that appear at the end of this book.

Throughout the work metres and kilometres are employed for heights and distances, and metres per second for wind velocities. The direction of the wind is given in degrees from the North point, so that an East wind is 90°, a South wind 180° and so on. The use of metric units has been adopted because they are used by the International Commission for Scientific Aeronautics, it being of great importance that observers in different countries should use the same units. In the case of atmospheric pressure the English unit of inches of mercury has been retained because the information concerning the pressure distribution at the times of the ascents has been taken from the publications of the Meteorological Office.

The investigation of the upper air by means of pilot balloons is a somewhat lengthy process, and involves a considerable amount of tedious calculation, which can however be much lightened by the use of the slide rule and mechanical calculators. Apart from all other work the plotting of the trajectories of the 200 ascents has involved the solution of some 8000 triangles.

The investigations were undertaken and this book was written at the suggestion of Dr W. N. Shaw, F.R.S., to whom my grateful thanks are due for introducing me to a most interesting field of study and for his invaluable help both in the course of the investigations and in the writing of this volume. He also kindly supplied me with the diagrams of surface pressure in Figures 32, 33, 35, 37, 39, and 40. I must also express my indebtedness to Mr W. H. Dines, F.R.S., who has helped me in a number of ways; without his unfailing assistance I should hardly have begun researches on the upper atmosphere. For the preparation of the diagrams that appear at the end of the volume I have to thank Miss Humphreys, a member of the staff of the Meteorological Office.

C. J. P. C.

DITCHAM PARK,
  PETERSFIELD.
  2 *April*, 1912.

CHAPTER I

THE STRUCTURE OF THE ATMOSPHERE AS DISCLOSED BY THE OBSERVATIONS OF PILOT BALLOONS AT DITCHAM

THIS book gives the results of 200 observations of pilot balloons or ballons sondes. The process of observation consists in watching the progress of small balloons as they rise through the air and are carried along by the winds. The majority of the observations were made at Ditcham in Hampshire, on the southern slopes of the South Downs. Fourteen ascents in May, 1907, were made at Totland Bay in the Isle of Wight, and two, on July 1st, 1907, at Chobham Common.

From the observations the height and horizontal distance of the balloon are computed, generally from minute to minute, by methods which will be described and discussed.

The purpose of the investigation is to determine whether the wind in the upper air is the same in direction or velocity as that at the surface, and to form a numerical estimate of the deviations that are observed. If we regard the air as composed of a series of horizontal layers or strata each with its own special wind velocity and direction at the time of observation, it is evident that the balloon rising with approximately steady upward motion will be carried along first by the surface current, and subsequently by the currents in the successive layers. We may suppose without inaccuracy that very little time is lost in adjusting the velocity of the balloon to the flow of the air current in which it happens to find itself at any moment, because the whole mass of the balloon is quite trifling compared with the forces that would be required to hold it against a flowing current of air ; and we are therefore justified in assuming that the horizontal distance traversed in any minute represents the average velocity of the layer passed through by the balloon in its journey aloft during that minute. Thus the horizontal velocities of the air in successive layers may be regarded as given by the observations. The thickness of the layers traversed in successive minutes depends upon the rate at which the balloon is ascending. In many cases this has been assumed to be uniform, and reasons will be given for considering that this assumption is sufficiently nearly justified for the results to be relied upon as giving at least a general representation of the true motion of the atmosphere in successive layers.

For the sake of uniformity the heights in the atmosphere will be given in kilometres, and for general purposes we shall consider the atmosphere as made up of a pile of layers each a kilometre in thickness. The highest of the observations deal with a pile of 18 such kilometre layers, which took 80 minutes or more to observe, but the large majority were much less thick. In some ascents only one or two kilometre layers were traversed, but a large number included five kilometre layers.

On each occasion the balloon was watched until one or other of the following events happened; either the balloon became invisible because it entered the clouds, or it was seen to burst and begin a precipitate descent, or it became so small that the observer was no longer able to identify the speck in the telescope. Sometimes this last event happened in consequence of the gradual diminution of the speck beyond the power of the eye to identify it, and on other occasions the observer after taking his eye from the telescope could not again find the speck. Sometimes in hazy weather the balloon was lost to sight when, had it been clear, its diameter would have been easily discernible.

A large number of the ascents were made in the evening, as at that time the convection movements due to the heating of the ground by the sun's rays no longer interfere much with the uniform ascent of the balloon; moreover the haze, which was found to interfere with the observations during cloudless days, was less at that time than at an earlier hour; it was also found that near the time of sunset, when the sky was becoming less bright, the balloon illuminated by the sun's rays became very easy to see at long distances; in some cases a balloon was watched for many minutes after the sun had set on the earth's surface, the balloon shining brightly and looking like a planet seen through a telescope; on one occasion a balloon was seen to burst under such circumstances at a horizontal distance of about 40 miles. The time of sunset was also convenient because it corresponded nearly with one of the hours of observation for the Daily Weather Report of the Meteorological Office.

The heights reached, and in many cases the immediate cause of the termination of the experiment, are given in the general table of results.

It will be understood from this description of the mode of procedure that from each ascent one can form a mental picture of the successive currents of air traversed by the balloon in its journey. It is true that the geographical position of the balloon will be different for each minute, but these differences are so small compared with the extent of the atmospheric current that the variation in position may be disregarded; and as a general rule we may regard the position of the successive currents as applicable to the atmosphere immediately above the observer at the time of the commencement of the ascent.

The mental pictures thus obtained of the succession of air currents which constitute the structure of the atmosphere over the observing station are of the most varied and sometimes of the most complicated character. In order to enable the reader to carry with him an idea of the sort of structure which may be disclosed by the observations of a pilot balloon the various features have been classified according to some prominent characteristic which is easily recognised in the diagrams representing the results of the ascents.

A further note must be made before proceeding to consider the different types of structure which have been disclosed by the observations. The wind close to the surface is influenced by the shape of the ground and other obstacles from which the upper layers are free. Hence the variations close to the surface are often of a specially complicated character having little relation to the structure of the atmosphere as a whole. Generally speaking the wind increases from its surface value in direct proportion to the height above sea level (see Chapter IX), with some little veer in direction until a height of between half a kilometre and a kilometre is reached. Thereafter the effect of the surface may be regarded as no longer applicable. The surface wind is accordingly a very unsatisfactory datum to which to refer the variations in the upper air. Generally speaking the wind gradually approximates to that computed as the "gradient wind" (see page 32) from the distribution of pressure at the surface. In many ways the gradient wind is a better datum than the observed surface wind and it has been noted in the tables and marked on the diagrams whenever a reasonably satisfactory computation could be made. In some cases however the distribution of pressure in the neighbourhood of the station is too irregular and too ill-defined for a satisfactory computation of the gradient wind to be made.

With this explanation we proceed to refer to some of the principal types of structure of the atmosphere which a study of the diagrams has disclosed. It must be remembered that they are necessarily limited to the occasions when balloons can be followed with a telescope. These are generally occasions of clear weather. During rain, or when there is fog or low cloud, observations are not possible.

The figures that follow are taken from cardboard models prepared to show the distribution of wind direction and velocity with height. Each card shows by its direction and length the wind direction and velocity at each kilometre of height[1]. In general the light coloured cards represent winds from between 300° (W.N.W.) through North to 120° (E.S.E), that may be supposed to come from polar regions; the dark cards winds from 120° through South to 300°, that may be supposed to come from equatorial regions. This classification is only approximate, since it is evident that, for example, a Northerly wind at the station may be a current of air that has been drawn from an equatorial region, but has curved round and passes over the station from a Northerly direction.

(a) "Solid" current. The first characteristic type is that which we may call the "solid" current, that is to say that after the interference of the surface has been passed, and the gradient velocity approximately reached the wind remains steady both in direction and velocity in the upper layers. This case is illustrated by the diagram representing the result of the ascent on May 5th, 1909 (Fig. 1).

(b) Continued increase of velocity beyond that of the gradient wind. In some cases it is possible to explain the increase of wind velocity by the change in the distribution of pressure in the upper layers without any discontinuous change in the air supply by reference to the distribution of pressure and temperature on the surface. Often however no such explanation is evident. These cases are illustrated by the results of ascents on Sept. 1st, 1907 and Oct. 1st, 1908 (Figs. 2 and 3).

[1] The arrow-head flies with the wind.

(*c*) Decrease of velocity in the upper layers. In this case after the gradient velocity has been reached the velocity falls off showing that the regime indicated by the surface pressure is over, and a new distribution commences, arising so far as we know from causes unrelated to the surface conditions. Sometimes the new regime is itself indicated by the observations at higher levels; not infrequently the observations had come to an end before the conditions in the higher levels were disclosed. This class is illustrated by the result of the ascent on May 7th, 1909 (Fig. 4).

(*d*) Reversals, or great changes of direction in the upper layers. Reversals of

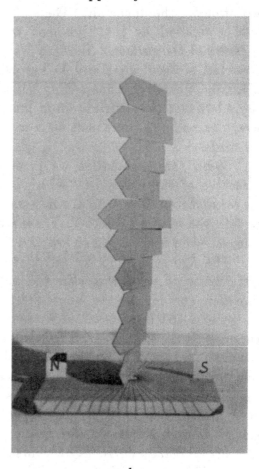

a   *b*

Fig. 1. Model representing the vertical wind distribution on May 5, 1909, 6.43 p.m. Class (*a*).

a. From South side.   *b*. From West side.

direction are nearly always preceded by the falling off of the velocity to a light air. These cases may therefore perhaps be regarded as a continuation of ascents similar to those of class (*c*) if it had been possible to continue the observations. In these cases we see the superposition of distinct systems of currents without any specific relation between them that can be accounted for by a knowledge of surface conditions. This class is illustrated by the result of the ascent of Nov. 6th, 1908 (Fig. 5).

(*e*) Upper wind blowing out from distant low pressure centre; frequent reversals in the lower layers. A number of cases present themselves in which the upper wind is completely at variance with the gradient wind, either in direction, in velocity, or in both. In many of these cases it seems as though the upper wind were blowing outwards from the region overlying a low pressure system. The wind in the upper layers is often far in excess of the gradient value, and increases as in class (b); there are cases in which the structure might be classified with either class. This class is illustrated by the result of the ascent of April 29th, 1908 (Fig. 6).

(f) The wind in the Stratosphere. In several ascents balloons have been followed to great heights and have been kept in sight after they have entered the region of the stratosphere. It has been found from the results of the ascent of ballons sondes that after a certain height is reached, a height that varies from day to day and from place to place, the ordinary diminution of temperature with height ceases, and that the column of air above this height remains at approximately the

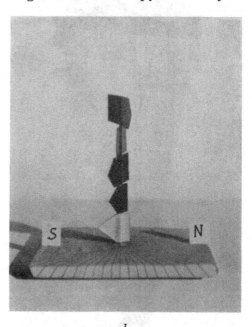

a   b

FIG. 2. Model representing the vertical wind distribution on Sept. 1, 1907, 10.22 a.m. Class (b).
a. From South side.   b. From East side.

same temperature as far as the observations have continued. This upper region of the atmosphere has been called the Stratosphere or Isothermal Layer. It has been found that at a height corresponding roughly with the commencement of the stratosphere the wind velocity usually decreases, sometimes in a very marked way. This class is illustrated by the results of ascents of July 29th and Oct. 1st, 1908 (Figs. 3 and 7). So far as observations of the movements of balloons are concerned the region thus identified is a region of lighter winds as compared with those of the strata beneath.

Mr W. H. Dines, F.R.S., informs me that he has had a few cases of no appreciable decrease in velocity up to 16 km.

It will be noticed that in many cases equatorial winds have polar winds above them. The frequency is too great for this to be merely coincidence, but it is quite possible that this superposition represents the condition of clear weather necessary for the observation of balloons. It may be argued that in the converse case when an equatorial wind is above a polar one we should get conditions favourable for the

a   *b*

FIG. 3. Model representing the vertical wind distribution on Oct. 1, 1908, 4.20 p.m. Class (*b*).
   a. From East side.   *b.* From South side.

formation of clouds and rain, and consequently conditions unfavourable for the observation of balloons, and some evidence will be brought forward to show that this is sometimes the case.

a

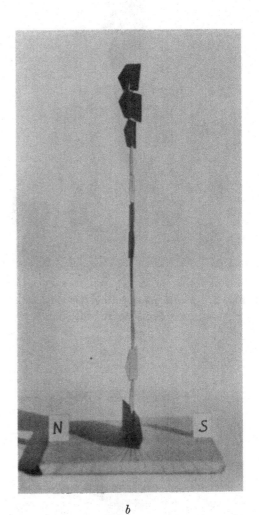

b

FIG. 4. Model representing the vertical wind distribution on May 7, 1909, 6.29 p.m. Class (*c*).
a. From South side.   *b*. From West side.

a   *b*

FIG. 5. Model representing the vertical wind distribution on Nov. 6, 1908, 10.59 a.m.   Class (*d*).
  a. From South side.   *b.* From East side.

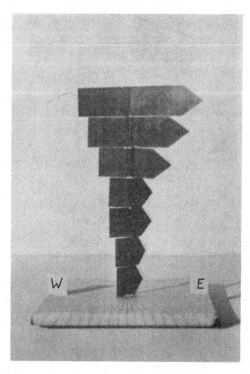 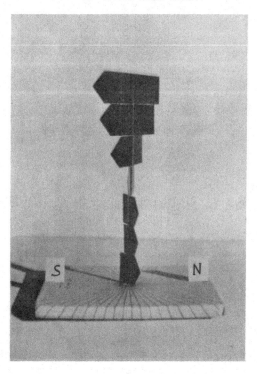

a   *b*

FIG. 6. Model representing the vertical wind distribution on April 29, 1908, 3.57 p.m.   Class (*e*).
  a. From South side.   *b.* From East side.

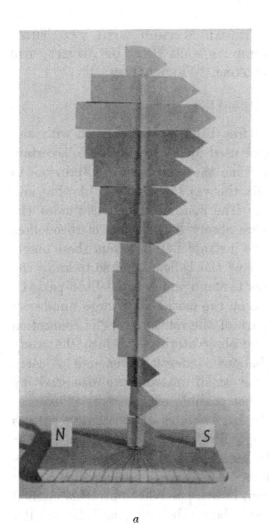

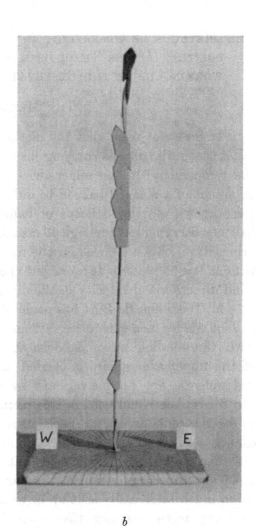

a   b

FIG. 7. Model representing the vertical wind distribution on July 29, 1908. Class (b).
    a. From West side.   b. From South side.

CHAPTER II

THE METHODS OF OBSERVING; (*a*) TRIGONOMETRICAL METHOD WITH TWO THEO-
DOLITES; (*b*) ONE THEODOLITE WITH ASSUMED UNIFORM RATE OF ASCENT; THE
WORKING UP OF THE OBSERVATIONS. BALLOONS. THEODOLITES

THE METHODS OF OBSERVING

IT does not seem that the motion of small free balloons was studied with any exactness until balloons carrying instruments were used by meteorologists to ascertain the temperature in the upper air. Before that time they had been used previous to an ascent of a manned balloon to indicate roughly the way the wind was blowing, and were known as pilot balloons or ballons d'essai. The motions of ballons sondes, the balloons carrying meteorological instruments, were observed by means of theodolites primarily to obtain a check on the readings of the instruments, but from these observations it was possible to work out the trajectory of the balloon, and so to know the wind direction and velocity in the different layers through which the balloon passed.

M. Teisserenc de Bort has made observations on the motion of a large number of ballons sondes, and such observations are now part of the routine work in connection with the sending up of balloons at most of the observatories at which the study of the upper atmosphere is carried on. But ballons sondes large enough to carry instruments are expensive, and even with the small instrument, designed by Mr W. H. Dines and used in this country, it is not possible on the score of expense to make ascents except on special occasions, and a large number of consecutive ascents cannot easily be carried out. Some years ago Professor Hergesell and other observers on the Continent used quite small balloons without instruments to determine the air currents above the earth's surface. The balloons being cheap, and being much more easily sent up in windy weather, it was possible to use this method on many more occasions than was the case with larger balloons, and far smaller quantities of hydrogen were necessary for their inflation.

The series of observations discussed in the following pages was begun at the close of the year 1906 and has been continued at intervals up to the present time.

Two methods of observing are in use (*a*) the two theodolite method, and (*b*) the one theodolite method.

(*a*) The two theodolite method.

A base line is selected, at each end of which is an observer with a theodolite. The base line should be as long as possible, but the balloon at one end must be clearly visible to the observer at the other. It is an advantage if each station is against the skyline as seen from the other, and particularly the home station from which the balloon is liberated should be seen against the skyline from the other station. In the cases discussed in these pages it was not possible to arrange this and several balloons were consequently lost to sight at the out station on days when the seeing was not good. The length of the base line employed was 2680 metres, but even in this country where atmospheric conditions are apt to be unfavourable a longer base might with advantage be used. The direction of the base line is also important; if possible it should be at right angles to the direction in which the balloon is likely to travel. A triangular base is a great advantage if it is possible to have observers at three stations, and use the stations to check each other; the balloon must travel in a direction favourable for observation from at any rate two of the stations. The observations carried on at Ditcham laboured under a disadvantage in respect of the direction of the base line; the only convenient base line was nearly north and south and as a considerable number of cases occurred in which the balloon travelled in a northerly or north-north-westerly direction it was not found possible to calculate the trajectories in these cases with any degree of accuracy from the two sets of observations.

The method of starting the balloons was as follows. The observer went to the out station and erected his theodolite; when he was ready he hoisted a flag and kept a watch on the home station. The observer at the home station as soon as he saw the flag at the out station brought out the balloon ready for the ascent. One minute before the commencement of the ascent the observer at the home station hoisted a flag, which was dropped at ten seconds before the start; as soon as the balloon was released the observer at the out station started a stop watch. At both stations readings of the bearing and altitude of the balloon were taken at each minute from the start.

The resulting observations were tabulated and dealt with as follows: AL (see Fig. 8) is the base line, B is the balloon and O a point vertically below it; the angles measured by the theodolites are the bearing OAL $(=a)$ and the altitude BAO $(=\beta)$; the bearing OLA $(=\lambda)$, and the altitude BLO $(=\gamma)$; these four angles were tabulated; in another column were tabulated the values of $a+\lambda$, $=\pi-\theta$; in working out the results a slide rule was used with the scale of sines and tangents uppermost; the value of $\sin(a+\lambda)$ on SS (see Fig. 9) was placed under the value of the base line on A or A', and over the value of $\sin a$ on SS was read the value of LO on A or A'; the right-hand end of the SS (or what is the same thing the TT) scale was then brought under the value of LO on A or A' and over the value of $\tan\gamma$ on the TT scale was read (by the aid of the cursor) the value of OB on the A or A' scale. If γ exceeds 45° the value of $(90°-\gamma)$ on TT is placed (by the aid of the cursor) under the value of LO on A or A' and over the right-hand end of the TT scale is read the value of OB on the A or A' scale. In the above cases it is convenient to number the angles on the slide rule to correspond with $a+\lambda$ on the sine scale and $90°-\gamma$ on the tangent scale; for instance if $a+\beta=110°$, $\sin 110°=\sin 70°$, and 110 is marked on the SS scale under 70; similarly when dealing with the tangent scale for angles

2—2

greater than 45°; if $\gamma = 60°$; we wish to multiply by tan 60°, that is to divide by tan $(90° - 60°) = $ tan 30°; we mark the slide rule with 60 under 30 on the tangent scale; if the slide rule is marked in this way a great deal of time is saved in working out the results.

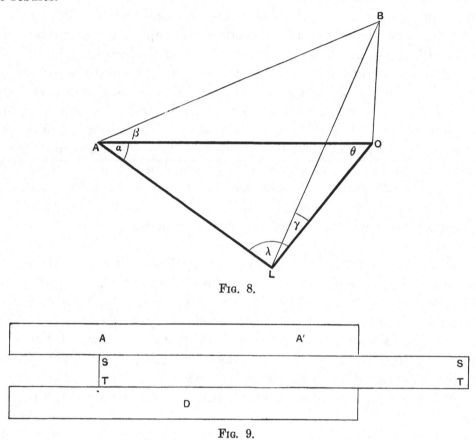

FIG. 8.

FIG. 9.

LO having been found and *OB* from it, *AO* may be found in a similar manner, and *OB* from this and from the altitude β; if *A* and *L* are at different heights allowance must be made for this in calculating *OB* (in figure 8 *A* and *L* are taken to be at the same height); after allowing for the difference of height the two values of *OB* thus found should agree; if they do not, some error has been made in the calculations or in the readings, or one of the theodolites is in error; if in a long series of observations the two values of *OB* differ in the same proportion it is generally a proof that there is something wrong in the setting or in the adjustments of one or both theodolites and this is a useful way of ascertaining such error.

It may happen that the trajectory of the balloon is so nearly in the same direction as the prolongation of the base line that the above method of calculating the heights and distances is no longer applicable. In such cases the following method has been employed.

A and *L* (see Fig. 10) are the two stations, *LA′* is the base, $AA′ = h$ is the height of *A* above *L*, and *b* is the length of the base line *LA′*.

Now
$$AO = OB \cot \beta,$$
$$LO = LA' + AO = O'B \cot \gamma,$$
$$LA' = OB (\cot \gamma - \cot \beta) + h \cot \gamma,$$
$$OB = \frac{b - h \cot \gamma}{\cot \gamma - \cot \beta} = \frac{\sin \beta}{\sin (\beta - \gamma)} (b \sin \gamma - h \cos \gamma)\dots\dots\dots\dots(1).$$

If O', A', and L are not in line draw $A'a'$ (see Fig. 11) at right angles to LO; a' may be considered as a point below an imaginary station a in the vertical plane through L and O; the base La' is given by

$$La' = LA' \cos \lambda = b \cos \lambda \ \dots\dots\dots\dots\dots\dots\dots\dots(2),$$

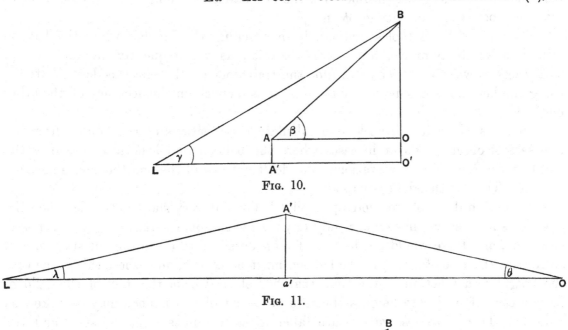

FIG. 10.

FIG. 11.

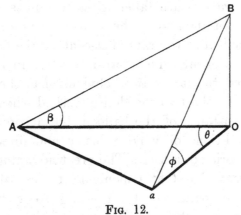

FIG. 12.

ϕ, the altitude from a is given by
$$AO \tan \beta = aO \tan \phi \quad \text{(see Fig. 12)},$$
and
$$aO = AO \cos \theta,$$
$$AO \tan \phi \cos \theta = AO \tan \beta,$$
or
$$\tan \phi = \tan \beta \sec \theta \ \dots\dots\dots\dots\dots\dots\dots\dots(3).$$

Putting ϕ for β, and $b \cos \lambda$ for b in (1) we get

$$OB = \frac{\sin \phi}{\sin (\phi - \gamma)} (b \cos \lambda \sin \gamma - h \cos \gamma) \quad \dotsc\dotsc\dotsc\dotsc\dotsc\dotsc(4).$$

With the aid of a Brunsviga calculator a table was constructed showing what should be added to β to give ϕ for values of β between 5° and 60°, and for values of θ from 1° to 16°; another table was constructed giving the numerical value of $(b \cos \lambda \sin \gamma - h \cos \gamma)$ where $b = 2680$ metres and $h = 105$ metres for values of γ from 5° to 25° and for values of λ from 0° to 10°.

With the aid of a slide rule it was then possible to calculate the height of the balloon when it was nearly in line with the two stations as long as there was sufficient difference between the altitudes ϕ and γ.

The use of these tables was a help in checking the heights when the balloon went in a Northerly or North Westerly direction, as was frequently the case; but in such cases it was found better to plot the trajectory on the one theodolite method; using the heights as calculated by the tables as a check on the accuracy of the other method.

In cases where it was useless to calculate the positions of the balloon from the two sets of observations, or in cases where the balloon had been lost at one of the stations, the remaining observations were dealt with as in the one theodolite method.

(b) The one theodolite method.

It is often desirable to send up a pilot balloon at very short notice, as when the sky clears for a short time on a cloudy day; or when some particular phenomenon is occurring, and there would not be time for an observer to go to the out station.   In such cases the rate of ascent of the balloon must be known, and where this is the case the height is a function of the time that has elapsed from the start of the balloon. In the case of moderate heights the rate of ascent of the balloon may be taken as uniform.   Theodolite observations are taken at each minute from the start and the position of the balloon is obtained from the bearing, the angular altitude and the assumed height.   The problem of the rate of ascent of the balloon, and of the effect of rising or falling currents of air, which certainly occur in the lowest strata of the atmosphere will be referred to later.   It is obvious that the one theodolite cannot give such accurate results as the two theodolite method when the base line is fairly large compared with the distance of the balloon, but it probably gives a very fair approximation to the conditions of wind velocity and direction; with respect to direction it is probably superior to a kite.   The great advantage of the one theodolite method is that it can be used almost at a moment's notice, whereas the two theodolite method requires more preparation; the working up of the results of the observations is far less laborious, and three or four sets of observations could be worked up for the one theodolite method in the same time that one could be done from the double observations.   Evidence will be given later to show that when the balloon is in a bad direction for observation or when it gets fairly distant with respect to the length of the base line the one theodolite method is actually superior to the two theodolite method.

In the case of either of the methods of observing we obtain a series of positions which gives the horizontal trajectory of the balloon. This is drawn out to scale (Fig. 13) and from the length of the minute runs is measured the wind velocity during that minute; that is assumed to be the wind velocity corresponding to the mean height of the balloon during that minute. The wind velocities are read off the diagrams by means of a scale which gives the values of the wind velocities directly.

Most of the diagrams have been drawn on a scale of 1 to 30480. This scale,

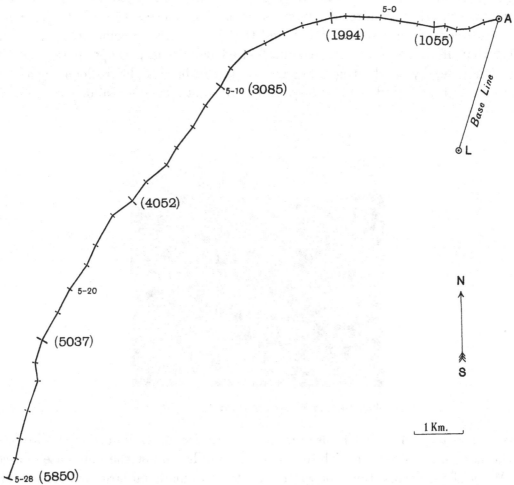

Fig. 13. Trajectory of pilot balloon, Feb. 22, 1909, 4.52 p.m. Observed with two theodolites. The cross lines show the position of the balloon at each minute; the figures in brackets are the heights in metres above the sea level.

which is 1000 feet to 1 centimetre, was adopted in the first instance because English units were used in the measurements but the diagrams were drawn on millimetre paper; with the use of metric units a more convenient scale might be found, but as a number of diagrams had been made before this change was made in the units it was thought better to keep the same scale; it offers no difficulties except that a special scale had to be made to measure off kilometres on the diagrams. If the run of the balloon is very long the trajectory is drawn out on half the above scale.

The wind directions are read off the trajectory by means of a transparent celluloid square ruled as a protractor. The wind direction is given in degrees from the North point; East being 90, South 180, West 270 and North 360 or 0.

When the values of the wind velocity and direction have been read off and tabulated a diagram is constructed giving the relation between the height and the wind elements.

In the case of ordinary pilot balloon ascents the balloons used are rubber balloons weighing from 28 to 30 grammes (Fig. 14); they are filled to lift from 85 to 90 grammes; and their rate of ascent is about 152 metres (500 feet) per minute. They are coloured dark red as it is found that if they are so coloured they are more visible than if they are white. An uncoloured balloon is perhaps more easily seen against a blue sky at the beginning of the ascent but as the balloon expands the rubber gets thinner and more transparent. At first a smaller balloon was used, but

Fig. 14. Pilot balloon ready for an ascent.

it has been found convenient to use rather a larger size as stated above and always to use the same kind of balloon with the same lift in order to get the same rate of ascent.

Many of the observations for wind velocity were made on large balloons carrying instruments (ballons sondes) which are sent up from Ditcham on the days arranged by the International Commission for Scientific Aeronautics for combined observations. These balloons are much larger than the pilots and many different sizes have been used. The most usual size is one weighing about 250 grammes; they are filled with hydrogen to lift 200, 300 or more grammes; at the 300 gramme lift they have a diameter of nearly one metre.

The height attained before bursting both by the large balloons and the small ones varies greatly; it depends principally on the quality of the rubber, and on how the balloons have been kept. It is well not to keep them long as the rubber deteriorates rapidly. It is best to keep them in a tin case with tissue paper to prevent them from being in contact with the metal; the case should be kept in a warm place, and the balloon may be well and evenly warmed just before inflation. In any case they should be kept in the dark as light has a deteriorating effect on rubber.

While it is being filled the balloon is attached to a balance (Fig. 15). By means of this instrument the lift of the balloon can be balanced against weights, and in this manner the balloon can be given a lift which can be measured without detaching it from the pipe by which it is filled.

The observations were taken by the special theodolite designed by Dr de Quervain for observing balloons. Two such theodolites, Fig. 16, made by Herr Bosch of

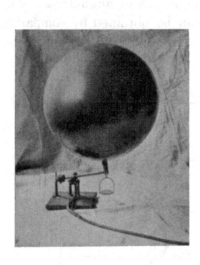

Fig. 15.   Balance used when filling balloons.   Fig. 16.   Theodolite for observing balloons (Bosch).

Strassburg were used until the spring of 1909. After that date most of the one theodolite observations were taken with a somewhat similar instrument made by Messrs Cary Porter. The telescope of these theodolites has a reflecting prism which reflects the light at right angles in such a way that the observer is always looking in a horizontal direction whatever the altitude of the balloon. The arm of the telescope that carries the eyepiece passes through the centre of the vertical circle, so that the observer has merely to take his eye from the eyepiece to read the vertical angle, instead of having to read it from the side as in the ordinary pattern of theodolite. The horizontal circle is also arranged so as to be easily read from the same position. The Bosch theodolites have a telescope with an object glass 5 cm. in diameter; with the eyepiece supplied the magnification is about 20 diameters. The Cary Porter instrument has an object glass of 6·1 cm. diameter, and has three eyepieces.

CHAPTER III

CHECKS ON THE ACCURACY OF THE METHODS OF REPRESENTING THE RESULTS OF THE OBSERVATIONS

CONSIDERABLE doubt has been thrown on the accuracy of the one theodolite method[1] and it is evident that if there are up and down movements in the atmosphere the values for wind velocity obtained from the observations of angular altitude will be more or less at fault.   A check on the method can be obtained by comparing the trajectory drawn by the two theodolite method with one drawn on the one theodolite

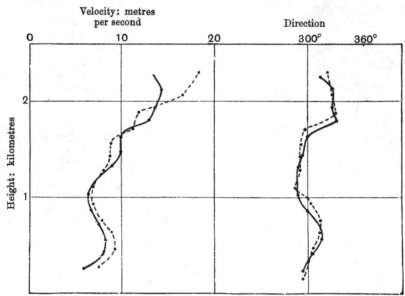

FIG. 17.   Feb. 26, 1908, 10.28 a.m.   Height-wind diagrams: comparison of one and two theodolite methods.

●———●———● One theodolite.
·———·———· Two theodolites.

method from one only of the sets of observations used in determining the trajectory in the first case.   Figs. 17 and 18 show the height-wind diagrams plotted from trajectories drawn out on both the method of one and of two theodolites.

The wind velocities were fairly large above two kilometres, and the balloon was rising faster than the normal rate at the end of the observations, possibly owing to an

[1] *The Free Atmosphere in the Region of the British Isles*, W. H. Dines, F.R.S., p. 28.

upward current caused by hills; up to two kilometres the agreement between the two diagrams is tolerably close.

Fig. 19 shows the height-wind diagram drawn out by both methods for an ascent on Feb. 18th, 1909 when there was a reversal at about 3·7 km. The two methods of plotting the trajectory give remarkably concordant results in this case; the wind directions are in disagreement to a certain extent in some cases, particularly when the velocities are small; this is only to be expected, for in the graphical method employed it is not easy to measure accurately the direction of a very short line; and a small error in the position of the points representing the positions of the balloon at two consecutive minutes causes a large error in the direction line when the points are close together.

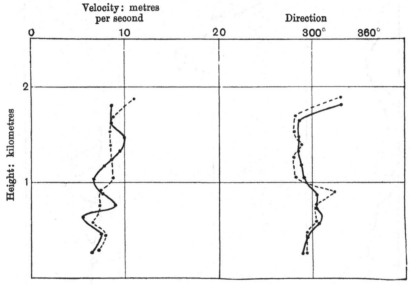

FIG. 18. Feb. 26, 1908, 11.5 a.m. Height-wind diagrams: comparison of one and two theodolite methods.

•————•————• One theodolite.
·— — —·— — —· Two theodolites.

Fig. 20 shows a pair of trajectories for June 3rd, 1908. It is I think clear from this diagram that when the balloon is at a great distance compared with the length of the base line the one theodolite method is the more accurate; the irregularities between the positions at 7.45 and at 7.54 are certainly due to small errors in the bearings; it is extremely unlikely that the wind between adjacent layers in the atmosphere should have fluctuations such as those shown on the two theodolite trajectory; they would also entail large fluctuations in the vertical velocity; a large vertical velocity being accompanied by a large horizontal velocity and *vice versa*; while in the case where the balloon appears to reverse its direction its height would have to decrease at the same time. The one theodolite method in this case is at any rate not open to these objections as will be seen from the figure.

When therefore the balloon gets into a bad position with regard to the base line, or when its distance becomes very great compared with the base, a combination of the one and two theodolite methods is advisable, a rate of ascent being adopted which best fits in with the heights deduced from the observations.

In the case of very high ascents which have been observed with two theodolites

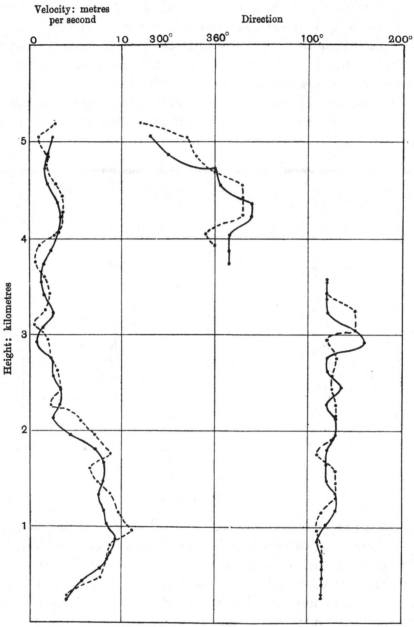

FIG. 19.   Feb. 18, 1909, 4.43 p.m.   Height-wind diagrams: comparison of one and two theodolite methods.

•———•———• One theodolite.
.———·———. Two theodolites.

it has been found that the assumption of a uniform ascensional velocity will not hold; a nearer approximation would be to suppose that the balloon ascended with a uniform acceleration; this assumption while not strictly true is probably sufficiently near the

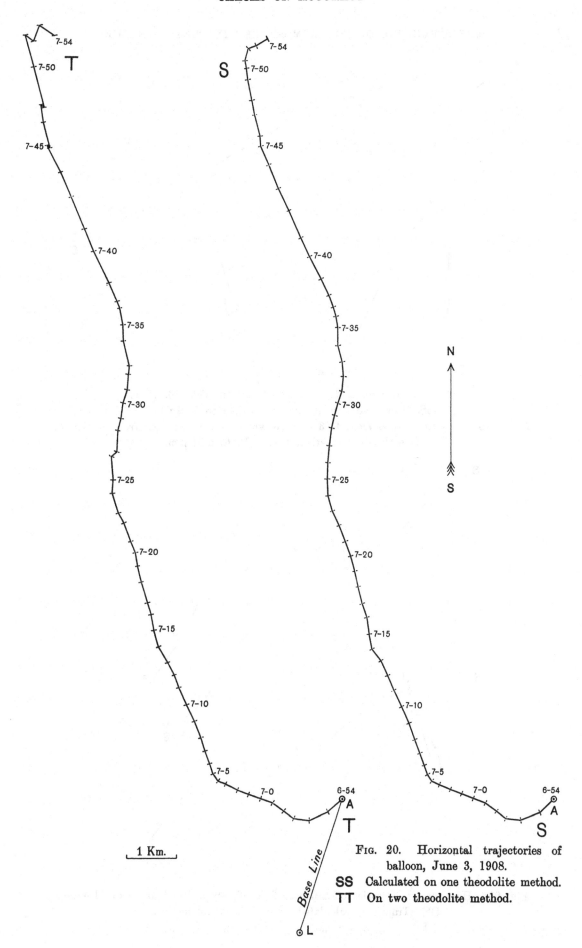

7-54
7-50 T

S
7-54
7-50

7-45

7-45

7-40

7-40

N

7-35

7-35

7-30

7-30

S

7-25

7-25

7-20

7-20

7-15

7-15

7-10

7-10

7-5

7-5

7-0

6-54
A
T

7-0

6-54
A
S

1 Km.

Base Line

L

FIG. 20. Horizontal trajectories of
balloon, June 3, 1908.
SS Calculated on one theodolite method.
TT On two theodolite method.

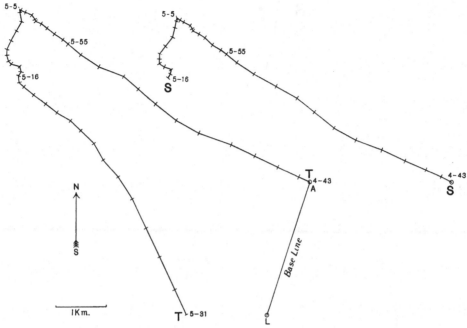

FIG. 21.   Horizontal trajectories of balloon, Feb. 18, 1909.
SS   One theodolite method from 4.43 to 5.16 p.m.
TT   Two theodolite method from 4.43 to 5.16 when balloon was lost from station **A.**
One theodolite method from 5.16 to 5.31 p.m.

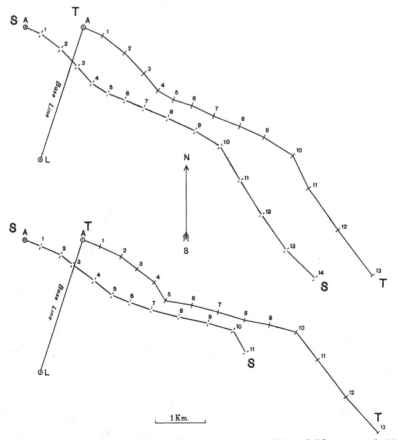

FIG. 22.   Horizontal trajectories of balloons, Feb. 26, 1908, 10.28 a.m. and 11.5 a.m.
SS   Trajectory calculated on one theodolite method.
TT   „   „   two   „   „

truth for the practical purpose of constructing diagrams giving the relation between
the height and the wind elements.   An acceleration is chosen which best fits in with
the heights as determined by the observations.   Fig. 21 shows part of a trajectory so
determined ; the balloon was at a horizontal distance of between 11 and 12 kilometres,
and at a height of between 12·5 and 15 kilometres during this part of its trajectory.
The two trajectories are in extremely close agreement in most places, and it is
interesting to see the loop just after 7.56 making its appearance in both diagrams.

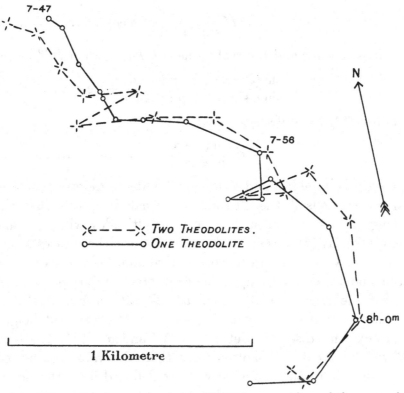

Fig. 23.   Part of trajectory of ballon sonde of Aug. 5, 1909 ; comparison of the one and two theodolite
methods.   The position of the balloon at 8.0 p.m. was 11·87 kilometres W. 13° S. of the starting point.

The one theodolite method has been criticised on the grounds that in particular
cases the rate of ascent of a balloon may differ
widely from the normal.   This may be the case
particularly when the balloon is near the ground.
If the balloon ascends with say double the normal
velocity it is evident that the real wind velocities
will be double those deduced from the assumed
and erroneous rate of ascent.   But it is uncommon
to find the rate of ascent differing widely from the
normal after the balloon has ascended a kilometre
or thereabouts above the surface.   Consider what
effect the increase or decrease of height, due to an
increase or decrease of ascensional velocity at the

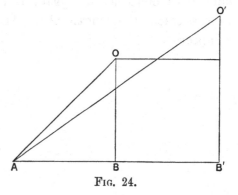

Fig. 24.

beginning of the ascent, will have if subsequently the balloon ascends with the normal velocity. Let A be the station (Fig. 24), O the position of the balloon after n minutes, and O' its position after $n+1$ minutes; the wind velocity is measured from the length of BB'; let $OB = H$ and $O'B' = H+h$ where h is the mean rate of ascent of the balloon; let the altitude of $O = a$ and of $O' = a'$.

Then
$$BB' = AB' - AB$$
$$= (H+h) \cot a' - H \cot a$$
$$= H \frac{\sin(a-a')}{\sin a \sin a'} + h \cot a' \quad\dots\dots\dots\dots\dots(1).$$

If the value of H is in error and the real value is $(H+d)$, then BB' becomes

$$H \frac{\sin(a-a')}{\sin a \sin a'} + d \frac{\sin(a-a')}{\sin a \sin a'} + h \cot a' \dots\dots\dots\dots(2);$$

subtracting (1) from (2) the difference is

$$d \frac{\sin(a-a')}{\sin a \sin a'} \quad\dots\dots\dots\dots\dots\dots\dots\dots\dots(3).$$

This expression is large if a and a' are small and if a differs considerably from a'; but in actual observations if the altitudes are small it means that the balloon is a long way off and in such a case the altitudes will not vary much between consecutive minutes; on the other hand in those cases where the altitude does vary much between consecutive minutes the balloon is near the station and a and a' are large; hence in practice during the later part of an ascent the expression (3) is small. To take a numerical case suppose the balloon to be 1000 metres higher than the assumed height (probably a very extreme case) and let $a = 15° \ 30'$ and $a' = 15°$; the true value of BB' will be about 126 metres more than the value calculated from the erroneous height, and the wind velocity deduced from the latter will be about two metres per second too large.

With a long base line the observed wind velocities for the first one or two minutes will be liable to error. If A and B be the stations (Fig. 25) and OA the horizontal trajectory the length OA will be proportional to $\sin OBA$; hence a small error in the angle OBA will cause a large error in the calculated length OA. The effect of this will be referred to in Chapter IX.

FIG. 25.

CHAPTER IV

The Rate of Ascent of Balloons theoretical and observed; Leakage of Gas from Balloons; the Relation of the Vertical Motion of Balloons to the Ground Contours

The Rate of Ascent of Balloons

The rate of ascent of balloons in still air is important in connection with the one theodolite method. With a rubber balloon[1] the free lift remains constant except in so far as it is affected by the tension of the rubber which increases the pressure inside the balloon to some extent. When the balloon has attained a steady rate of ascent, which it does in a few seconds after the start, the resistance of the air must equal the free lift and the rate of ascent is given by the equation

$$L = k\rho r^2 v^2$$

where k is a constant, ρ the density of the air, r the radius and L the free lift of the balloon. If ρ_1 is the density of the gas inside the balloon $\frac{4}{3}\pi\rho_1 r^3$ is its mass which is constant, except for leakage. $\frac{\rho_1}{\rho}$ may be taken as constant and therefore ρ varies as $\frac{1}{r^3}$, and as L is constant $\frac{v^2}{r}$ or $v^2\rho^{\frac{1}{3}}$ is constant, and v varies as $\rho^{-\frac{1}{6}}$.

Mr Dines gives the following table showing approximately the corresponding values for height, density, and ascensional velocity, taking into account the fall of temperature with height:

Height	ρ	v
Surface	1·00	1·00
3000 m.	·73	1·05
6000 m.	·53	1·10
9000 m.	·39	1·17
12000 m.	·26	1·25

In the ascents considered in these pages there were three in which the balloon remained in a direction and at a distance favourable for calculation by the two theodolite method, and in which the balloon was observed from both stations up to heights exceeding 12 kilometres. The calculations were made by combining the

---

[1] *The Free Atmosphere in the Region of the British Isles*, W. H. Dines, F.R.S., p. 27.

methods of one and two theodolites, as explained in Chapter II, and the following values were obtained :

Vertical velocity of balloon at different Heights.

	v at 4 km.	v at 12 km.	$\dfrac{v \text{ at } 12}{v \text{ at } 4}$	j metres per min.
Dines value	—	—	1·168	—
1908 Oct. 1	186·5	222	1·185	0·91
„ „ 2	202	253	1·253	1·52
1909 Aug. 5	152	188	1·237	0·75

Professor Hergesell has made a large number of observations of the rate of ascent of pilot balloons, both in the free air, and in the Cathedral and in the library of the University at Strassburg[1]. He finds that the rate of ascent of small rubber balloons is a function of the weight and the free lift of the balloon. If A is the free lift and B the weight in kilogrammes, Q the cross section is proportional to $(A+B)^{\frac{2}{3}}$. If R is the air resistance, $R = A = \phi Q . f(V)$ where V is the rate of ascent, and ϕ and f are functions to be determined.

Therefore
$$f(V) = \frac{A}{\phi Q}$$

or
$$V = F\left\{\frac{A}{\phi (A+B)^{\frac{2}{3}}}\right\}.$$

From the results of observations Professor Hergesell gives an empirical formula
$$(\phi) = q - \epsilon q^{*}$$
$$= (A+B)^{\frac{2}{3}} - \epsilon (A+B)^{\frac{4}{3}},$$

and he finds that $\epsilon = 0\cdot 8$ gives good agreement with observation.

Therefore
$$V = \frac{A}{(A+B)^{\frac{2}{3}} - 0\cdot 8 (A+B)^{\frac{4}{3}}},$$

for the values of A and B considered. A diagram is given showing the rate of ascent of balloons up to 200 grammes weight, and 250 grammes free lift.

In the observations at Ditcham no exact record was kept of the free lift or of the weight of the balloons used, but with the exception of a few used in the early ascents nearly all the pilot balloons weighed about 30 grammes and were given a free lift of about 85 grammes ; from Professor Hergesell's table this would correspond to a rate of ascent of 150 metres per minute. The rate of ascent which had been deduced from some rough observations in a closed space, and confirmed afterwards by two theodolite observations was 152 metres per minute ; this is in fair accordance with Professor Hergesell's subsequently published value.

[1] *Sixième Réunion de la Commission Internationale pour l'Aérostation Scientifique* (Schauberg, Strassburg), pp. 86 et seq.

* loc. cit. p. 98

Professor Hergesell made a number of observations of the rate of ascent of balloons using two and sometimes three theodolites. He gives a table showing the differences between the observed rates and the rate as calculated from the above mentioned formula, the difference being due presumably to an upward or downward current of air. He finds that at Strassburg there is usually an upward current of nearly 30 metres per minute in the first half kilometre; this upward current falls off with height and ceases altogether at about 4 kilometres. The following are the figures given from 13 ascents in which the balloon was observed up to 4 kilometres. The figures are the differences between the observed and the calculated rates of ascent between ground level and 0·5 km., ground level and 1 km. &c.

0–0·5	0–1·0	0–2·0	0–3·0	0–4·0
29·0	20·1	12·7	6·8	2·8

Above 4 kilometres there are signs of a descending current but the number of ascents are few and Professor Hergesell surmises that the observed effect may be due to a leak in the balloon.

The following are similar figures for ascents at Ditcham, the figures in brackets giving the number of ascents:

0–0·5	0–1·0	0–2·0	0–3·0	0–4·0
33·2	19·5	14·9	11·8	9·3
(31)	(31)	(24)	(11)	(5)

For the lower strata these figures agree fairly well with those found by Professor Hergesell, but for the higher strata the Ditcham values are higher than those from Strassburg.

Mr A. Mallock, F.R.S., has given the following formulae for the ascent of rubber balloons[1].

If W is the weight of the balloon and fittings and F the total lift, when the pressure outside is p_0, and F_0 when it is $p = p_0/m$; also let $e\ (= p_0/n)$ be that part of the pressure inside the balloon due to its elasticity, so that the actual pressure inside (assuming that e does not vary with height, which is not strictly true) is $p + \dfrac{p_0}{n}$ and $1/\kappa$ the ratio of the densities of the gas in the balloon and of the atmosphere at the same pressure ($\kappa = 16$ for hydrogen). If u_0 is the volume of the balloon at pressure p_0 and density ρ_0

$$F = g u_0 \rho_0 \left(\frac{p}{p+e} - \frac{1}{\kappa} \right); \quad F_0 = g u_0 \rho_0 \left(1 - \frac{p_0 + e}{\kappa p_0} \right),$$

therefore

$$\frac{F}{F_0} = \frac{n}{n+m} \frac{n(\kappa - 1) + m}{n(\kappa - 1) - 1} = M, \text{ say.}$$

Also if v is the velocity with which the balloon would rise from the ground if it had no weight, and did not compress the gas by its elasticity (i.e. if $n = \infty$ and $W = 0$)

[1] *Proc. Roy. Soc.* A. Vol. LXXX. p. 530.

4—2

the ratio of the upward accelerations at pressures p and p_0 is $\dfrac{F-W}{F_0-W}$ or $\left(\text{if } \dfrac{W}{F_0}=A\right.$ $\dfrac{M-A}{1-A}$ which must be equal to the ratio of the resistances experienced, viz.

$$\frac{\rho r^2 v^2}{\rho_0 r_0^2 v_0^2}.$$

Now

$$\frac{\rho}{\rho_0}=\frac{1}{m} \text{ and } \frac{r^2}{r_0^2}=\frac{1}{m^{\frac{1}{3}}}\left(\frac{n+1}{m+n}\right)^{\frac{2}{3}},$$

hence

$$\frac{v}{v_0}=m^{\frac{1}{6}}\left(\frac{m+n}{m+1}\right)^{\frac{1}{3}}\left(\frac{M-A}{1-A}\right)^{\frac{1}{2}}.$$

Thus the velocity of the balloon at first increases as the one-sixth power of the ratio of the density of the air at the elevation attained to the density at ground level and when n is large (that is when the elastic compression is small) the upward velocity reaches its maximum not far from the greatest elevation to which the balloon can attain.

The balloon will cease to ascend when $F = W$ and this leads to the following expression for the limiting value of m:

$$m=\frac{n^2(\kappa-1)(1-A)+nA}{A\{n(\kappa-1)-1\}+n}.$$

The pressure is then $\dfrac{760}{m}$ mm.

The balloon is of course losing hydrogen as it ascends. A balloon of the quality used for the majority of the ascents and filled to lift 85 grammes loses 27 % of its free lift in four hours; or at the end of this time its free lift would be about 71 grammes. According to Professor Hergesell's formula it would then have a rate of ascent of 140 metres per minute instead of about 150. The loss of buoyancy in respect of time is practically linear for four hours, and the diminution of the rate of ascent in one hour is negligible. But it must be remembered that the balloon is expanding as it ascends and the loss of gas would be greater as the balloon's diameter increased.

Connected with the subject of the rate of ascent of pilot balloons is the question of vertical currents in the atmosphere due to the wind blowing over irregularities on the ground. It is obvious that a hill would produce an upward current in a wind blowing against it, but what the amount of such an upward current would be has not been accurately determined. Captain C. H. Ley made a number of observations of pilot balloons, measuring the angular diameter of the balloons to ascertain the distance. From observations made in Herefordshire[1] he came to the conclusion that inequalities in the ground caused a marked effect in the upper air, and that rising currents due to hills were transmitted upwards to great heights, to as much as 20,000 feet in some cases. In a later paper however Captain Ley says that the

[1] *Quart. Journal R. Met. Soc.* Vol. XXXIV. p. 27.

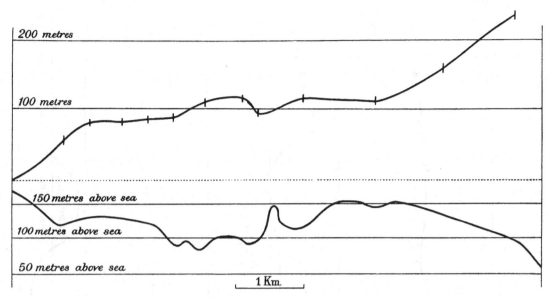

Fig. 26.   Relation of height of balloon to ground contours, Feb. 26, 1908.

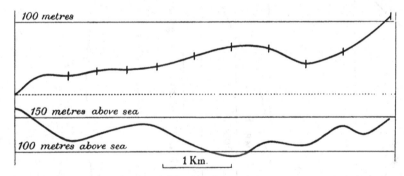

Fig. 27.   Relation of height of balloon to ground contours, Feb. 26, 1908.

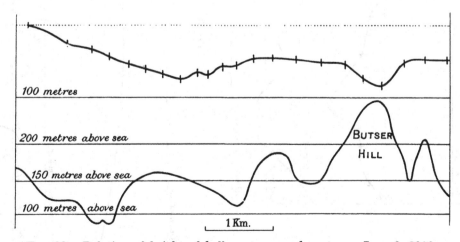

Fig. 28.   Relation of height of balloon to ground contours, June 3, 1908.

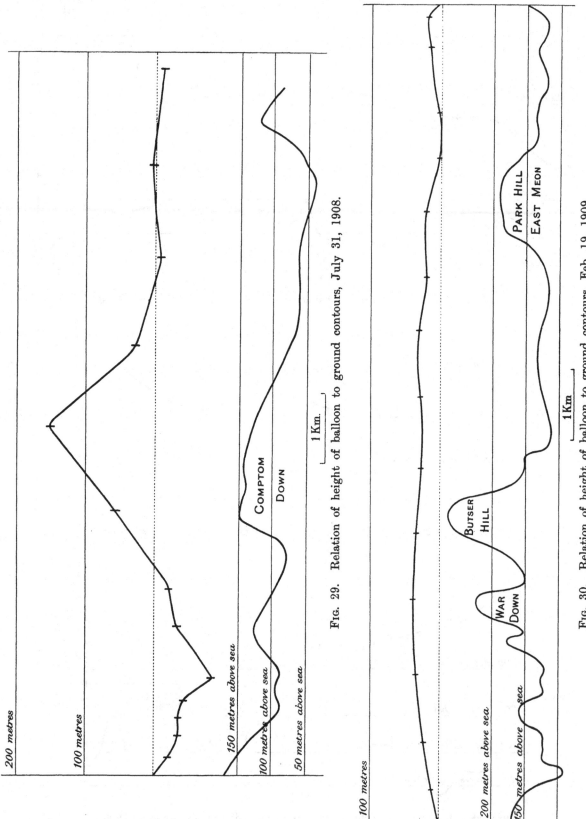

FIG. 29.   Relation of height of balloon to ground contours, July 31, 1908.

FIG. 30.   Relation of height of balloon to ground contours, Feb. 19, 1909.

effects of hills are "noticeable up to 1500 feet or more[1]." Balloonists have noticed an increased vertical velocity when going over hills especially when near the ground.

In order to examine these effects the figures 26 to 30 have been prepared; they show the height of the balloon in relation to the height of the ground contours; the dotted horizontal line represents the height at which the balloon would have been had it ascended with uniform velocity; the other line shows the distance of the balloon above or below this point for each minute from the beginning of the ascent. On Feb. 26th, 1908 there was a considerable rise as the balloon passed over a hill; there was a strong wind which was blowing nearly at right angles to Compton Down, a long hill, with a steep face on what was the windward side at the time. There were two balloon ascents, one closely following the other, and both show the rise on passing over the hill, both balloons were about 2 kilometres high at the end of the ascent[2]. On June 3rd the height of the balloon appears to have little connection with the ground contours; on this occasion the wind was very light and there are indications of a downward current between two thunderstorms. On July 31st, 1908 there was a rise over a hill and a drop in the vertical velocity behind it. On Feb. 19th, 1909 with a fairly strong wind a little above the surface the ground contours had no effect on the height of the balloon.

Further observations on this point would be useful, but I do not consider that the effect of ground contours would vitiate the results of one theodolite observations, though in the first one or two kilometres it might cause some errors in the calculated wind velocities.

It is known that considerable upward currents take place under cumulus clouds, but as most of the pilot balloon ascents here considered took place either in clear weather or with uniform sheets of cloud, there is no evidence on this point from these observations.

In the ascent at 6.54 p.m. on June 3rd, 1908 the balloon ascended with considerably less than the normal velocity; there is evidence of a downward current of 10 metres per minute between ground level and 0·5 kilometre; it is of course possible that the balloon was leaking and was therefore not ascending with normal velocity; at the time however there was a considerable quantity of false cirrus in the north, associated with a thunderstorm that had taken place in the afternoon, while to the south there was a large cumulo-nimbus in the distance, which produced a thunderstorm during the night; it seems likely therefore that between the two disturbances there was really a downward current which caused the balloon to ascend with a decreased velocity.

[1] *Quart. Journal R. Met. Soc.* Vol. XXXV. p. 18.
[2] Both the balloons unfortunately burst prematurely.

CHAPTER V

SUMMARY OF RESULTS AND THE RELATION OF THE WIND TO THE
SURFACE PRESSURE DISTRIBUTION

SUMMARY OF RESULTS

IN considering the 200 results obtained from the ascents it may be useful at the outset to have before us a note of what will be regarded as novel or unexpected, as distinguished from that which our previous knowledge would lead us to expect.

We may start from the consideration of the gradient wind as computed from the distribution of pressure. Recent work on the comparison of the observed wind at the surface, or in the lower sections of the upper air by means of kites, with the distribution of pressure at the surface have led to the general conclusion that the direction of the isobar is frequently an accurate representation of the direction of the surface wind. There is, however, usually a deviation of the surface wind from the isobaric line through some angle between 0° and 45° towards the low pressure side, and there are occasions of wider divergence which deserve further investigation. As regards velocity, we may calculate it from the measured distance of two isobars between which the station lies, by means of the formula

$$\gamma = 2\omega\rho V \sin \lambda$$

where γ is the gradient, ω the angular velocity of the earth, λ the latitude, V the velocity and ρ the density of the air.

This gives a result which is certainly related to the actual wind, but which may require some modification on account of the curvature of the path traversed by the air, and is in excess of the observed velocity at the surface on account of the friction between the moving air and the ground.

We cannot tell precisely what correction ought to be made for curvature of path because we do not know the curvature: we know however that if the curvature of the path is counter clockwise, as in a cyclonic depression, the velocity necessary to balance the gradient is less than that required for a straight path, and *vice versa* if the curvature of the path is clockwise like the isobar of an anticyclone then greater velocity is required to balance the gradient than would be required if the isobars were straight.

Neither do we know what correction ought to be made in any special case on account of friction. Our only means of dealing with such a question is a prolonged study of the relation of the observed surface wind to the gradient, which has not yet been undertaken. Consequently in comparing the surface wind with the gradient wind we must be prepared to make allowance for corrections on account of these known causes, before we come to consider whether the observations disclose other causes, or other evidences of departures hitherto unrecognised.

We know further that the distribution of pressure within the first kilometre of height is not likely to be very different in shape from that at the surface. A change in the distribution in height may be supposed to occur in consequence of the differences of density of columns of air over adjacent areas, due to differences of temperature, humidity or pressure, but for one kilometre these differences of density are small within a limited area. Hence as a first approximation (and doubtless with some exceptions) we may regard the distribution of winds indicated by the pressure at the surface as holding for the first kilometre or thereabouts. Above the surface the effect of friction will certainly be less marked, and in the upper air the comparison between the computed and actual winds will become freed from that influence. The agreement therefore should become closer above the ground level. This anticipation has been borne out by the observations recorded herein, and supported by others derived from kite ascents. From these we conclude that the gradient velocity will probably be reached within about one kilometre and the attainment of the gradient velocity marks the first stage in the structure of the atmosphere.

The transition from the surface velocity to the gradient velocity may also, as a first approximation, be regarded as a linear variation proportional to the height of the point above sea level (see Chapter IX).

We may therefore regard the first kilometre as the section within which the gradient velocity ought to be attained in normal cases. The cases in which the gradient velocity is not attained either as regards direction or velocity, or both, will be regarded as displaying some feature which should be the occasion of further investigation.

We have next to consider the part of the structure above the first kilometre. Here we have little to guide us. We cannot fairly assume that the gradient above is the same as at the surface, because the pressure distribution may be seriously affected by differences of density in different columns owing to differences in the temperature at the base, or to the temperature gradient of the column, or to other causes.

We are dependent therefore on the cases that actually present themselves, and we can only note that uniformity in direction and velocity of the wind above the first stage requires the lines of temperature and pressure to be similar and according to a certain law. Uniformity of direction requires isobars and isotherms to be parallel. Uniformity in velocity requires that there should be uniformity of temperature in each horizontal layer (i.e. distance between isotherms should be infinite). The surface gradient would certainly become intensified in the upper air if the lines of temperature distribution were similar to the lines of pressure distribution; it would be attenuated if the lines of temperature distribution were opposite to those of surface pressure, supposing in both cases that the vertical temperature gradients are identical.

In a few cases maps have been constructed to show the pressure distribution at some particular height above the surface, the pressures being calculated from the surface pressures and temperatures alone. It is obvious that such maps must be approximations only to the real pressure distributions at the heights considered; but there are cases where particular vertical wind distributions can be explained by supposing that the pressures in the upper air are due to the surface pressures and temperatures only. The difference of pressure at any height h above the stations s_1 and s_2 is given by the formula

$$p_2' - p_1' = p_2 - p_1 - \frac{t_1 - t_2}{491} \text{ (3 inches per kilometre)},$$

where p_2 and p_1 are the pressures and t_2 and t_1 the temperatures at the stations s_2 and s_1, at the ground, and p_2' and p_1' the pressures at the heights under consideration above the stations. The pressure above Portland Bill, the nearest station to Ditcham for which values are given in the Daily Weather Report, is taken as unity, and isobars are drawn for every tenth of an inch above and below the value at Portland Bill. Since the shape of the isobars is given by the relative pressures the numerical value of the pressures has not been calculated.

A reversal or a considerable deviation of the upper wind from the gradient wind, after it has once been reached, involves a change in the density conditions of the structure of the upper air which requires investigation.

Hence the different cases will be classified according to the transformation that takes place above the first kilometre, or wherever the nearest approximation to the gradient wind is reached.

It will be desirable here to indicate the method which has been adopted to represent the results. It will be remembered that the upper currents may show deviations from the surface current either in magnitude, or in direction, or in both, and in consequence we have to represent variation with height both in velocity and in direction. There is unfortunately no satisfactory method of representing these related variations in a single diagram. The most direct method of representing the results of an ascent is to make a plan of the path of the balloon as in Fig. 13, showing the trajectory for Feb. 22, 1909. In this case the orientation of the line shows the direction of the wind in each minute, and the distance of consecutive points marked on the diagram shows the magnitude of the velocity. The heights can also be written against each point, and thus all the information can be put on a single diagram, but it is a very extensive one, and the comparison of velocities by comparing distances does not easily present itself to the eye. We have therefore adopted the rather more recondite method of using two adjacent diagrams showing the variation with height of velocity and of direction respectively. A representation on this plan of the results of typical ascents, with weather maps showing the corresponding meteorological situation, follows the account of the general results.

As regards the direction of the upper currents it is a matter of ordinary experience that sounding balloons sent up under various conditions of weather are generally, though not by any means always, found at some point to the East of the starting point, and the idea of a dominant or persistent westerly current in the upper air forming part of the general circulation round the pole has been kept in mind, but the occasions upon which a westerly upper current has declared itself as characteristic of the highest strata reached are comparatively few.

The 200 ascents have been divided into the following classes :

(a) 1. "Solid" current; little change in velocity or direction; the wind reaches the gradient value and does not increase very much at greater altitudes.

2. No current up to great heights.

(b) Considerable increase of velocity; gradient value reached and surpassed; increase often accounted for by surface temperatures.

(c) Decrease in velocity in the upper layers.

(d) Reversals or great changes of direction.

(e) Upper wind blowing outward from centres of low pressure; frequently reversals at a lower layer.

1. Upper wind between West and North.

2. Upper wind between South and West.

Classes (a), (b) and (c) are represented diagrammatically in Fig. 31. There are many cases in which the wind does not reach the gradient velocity at all, but since many of these cases exhibit in other respects the same features as cases of classes (a) and (c) they have been included in these and will be dealt with under these classes. The consideration of the wind in the surface layer and in the Stratosphere will be dealt with in separate chapters.

There are a certain number of ascents which cannot be classed with any of the above. It must also be remembered that the ascent may terminate or the balloon be lost to sight at a low level when, had it been observed to a greater height, the ascent might have been included in another class.

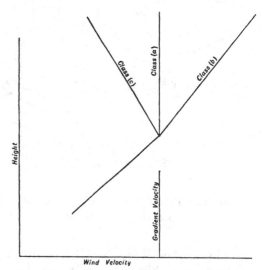

FIG. 31. Relation of wind velocity to height in Classes (a), (b), and (c).

Class (a). "Solid" current.

In this class (34 cases), the wind remains the same both in direction and in velocity in the upper layers. In order that this condition may be fulfilled the gradient and the shape of the isobars must be the same in the upper air as at the surface, and for this to be the case the temperature distribution on the surface must be fairly uniform, and this is found to be the case in the examples under consideration. There are some exceptions, Feb. 1st, 1907, and Feb. 20th to 24th, 1909, for example, when there was a definite temperature gradient, high temperatures being found to the West and low to the East; a map of the isobars at 2 kilometres on Feb. 1st, 1907 (Fig. 32) shows a gradient for northerly winds, but the gradient is not steep enough to cause any great increase in the wind velocity. A map of the isobars at 3 kilometres for Feb. 20th, 1909 (Fig. 33), shows a pressure gradient less than that at the surface; on this occasion the gradient velocity was not reached, the wind remaining uniform at a velocity slightly below the gradient velocity.

In most of the cases of this class the temperature distribution is either very uniform or irregular with no definite gradient.

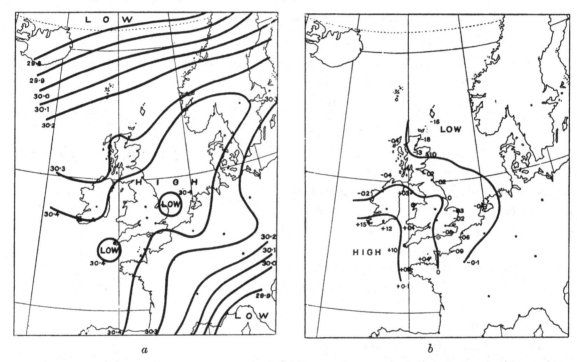

FIG. 32. Feb. 1st, 1907, 6 p.m. *a*. Surface isobars. *b*. Computed isobars at 2 kilometres, showing relative differences of pressure from that over Portland Bill.

The direction of the wind, both gradient and observed, has been tabulated for each case (see Fig. 34)[1]. It will be noted that there is no recorded surface wind between 135° and 235°; there are several cases in which the gradient wind is within these limits; in five of these cases the surface wind is backed and in one case it is veered with regard to the gradient wind direction. It seems probable that the absence of surface winds from the South in this class is really due to the configuration of the ground at the station; the line of the South Downs runs from East to West;

[1] The two following cases have not been included in the diagram: June 23, 1908, gradient direction 330°, surface direction 60°; April 17th, 1907, gradient direction 40°, surface direction 355°.

one would expect a wind from the East of South to be deflected more to the East and a wind from the West of South to be deflected more to the West on the southern slopes of the Downs. The deflection to the East is more marked than the deflection to the West, as should theoretically be the case ; an upper wind should be veered in respect to a surface wind owing to the Earth's rotation, apart from the local configuration of the ground.

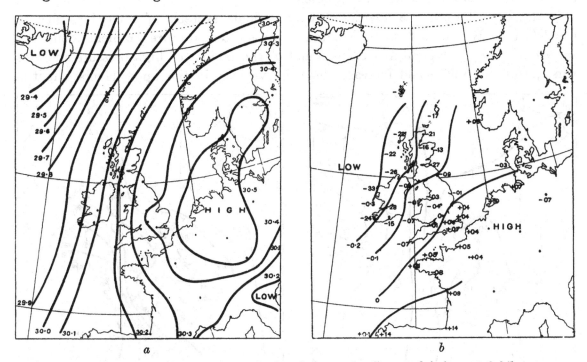

FIG. 33. Feb. 20th, 1909, 6 p.m.   *a.* Surface isobars.   *b.* Computed isobars at 3 kilometres showing relative differences of pressure from that over Portland Bill.

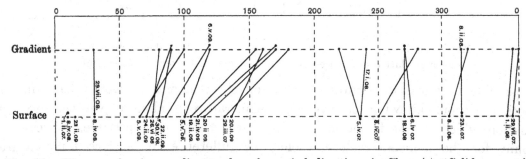

FIG. 34. Diagram showing gradient and surface wind directions in Class (*a*) "Solid current."

The cases of this class do not seem to be associated with any particular configuration of isobars.

Those cases when there is no wind to speak of up to great heights may be considered to belong to a subclass of case (*a*). The cases sometimes occur in anticyclonic weather, and what little wind there may be varies from height to height, so that winds of various directions are found superposed one on the other. The three ascents of Aug. 5th, 1909, are typical of this class which is an uncommon one.

Class (*b*). Considerable increase of velocity.

In this class the wind velocity increases rapidly with height; it usually reaches the gradient velocity at 0·5 to 1 kilometre, and the increase goes on rapidly to double the gradient velocity or even more.   April 2nd, 1907, is a typical case; the gradient velocity of 10 metres per second is reached at 0·5 kilometre; while at 1·6 kilometre the observed velocity is double the gradient.

The typical cases are associated with cyclonic systems of some intensity to the West or North, and there is often a strong temperature gradient along the surface, the high temperatures being found in the regions of high pressure and vice versa; this in itself would increase the wind velocity in the upper layers, but it is

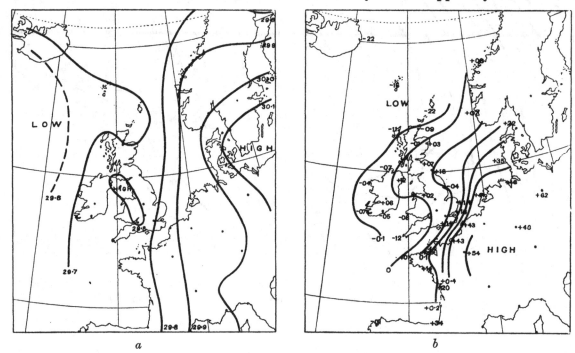

FIG. 35.   May 11th, 1907, 6 p.m.   *a.* Surface isobars.   *b.* Computed isobars at 3 kilometres, showing relative differences of pressure from that over Portland Bill.

not possible to calculate accurately what effects temperature would have in this direction without knowing the vertical temperature gradients over the area, information which is not forthcoming.   In many cases it is possible, however, to account for the increased velocity solely by differences of surface temperatures. May 11th, 1907, Fig. 35, is a case in point; the surface pressures are rather irregular with a gradient for light winds; a map of the isobars at 3 kilometres shows a very strong pressure gradient over the English Channel which quite accounts for the strong wind at that height.

Several cases are included in this class which differ markedly from the typical cases, as for instance May 18th and Feb. 5th, 1908.   In these cases a high pressure area is over or close to the station and there is a strong temperature gradient in the right direction for increased velocity in the upper air.

In comparing the gradient and the observed surface wind directions (see Fig. 36) it will be seen that the majority of cases of this class are from a direction between 135° and 180°, and from 270° to 360°; in other words the case is associated with advancing depressions or with depressions that have passed to the northward of the station.

The phenomenon of a gradient wind from a direction East of South becoming a surface wind from a more easterly point, and a gradient wind from a point West of South becoming a surface wind from a more Westerly point is again brought out in this class.

One or two cases require special mention. On Feb. 2nd, 1908, with a northerly wind at the surface, the wind at 4 kilometres reached a velocity of 30 metres per second, or three and a half times the gradient velocity. A V-shaped depression appeared over these islands on the following day, the main depression moving towards Scandinavia. A map of the isobars at a height of 3 kilometres, Fig. 37, shows that the gradient for northerly winds at that height is considerably steeper than on the surface.

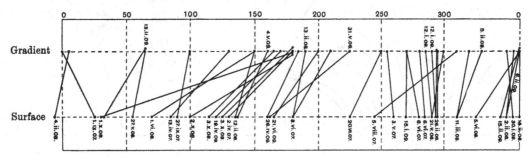

Fig. 36.   Diagram showing gradient and surface wind directions in Class (b).   Increasing velocity.

On Jan. 12th, 1909, is another case of extremely rapid increase in wind velocity with height; the temperature gradient is in the right direction for an increase of westerly winds in the upper air; it should be noticed in this case, however, that a deep depression moved down to the neighbourhood of this country by Jan. 14th and the case therefore has some resemblance to those in class (e).

Class (c).   Decrease in velocity in the upper layers.

This class is almost entirely connected with easterly winds on the surface. In a typical case, such as Jan. 2nd, 1908, the surface wind is high but considerably below the gradient velocity; the wind increases rapidly with height, often quite as rapidly as in Class (b); a maximum is reached at about 1 kilometre, above which there is a rapid falling off; the maximum, which is usually very well marked, sometimes exceeds and sometimes falls short of the gradient velocity; the latter is usually high, being often 20 metres per second or more.

The diagram (Fig. 38) showing the wind direction, both gradient and observed, at the surface shows the easterly character of winds that fall off in the higher layers.

With few exceptions the gradient winds fall between 50° and 150°, and the observed winds fall between 0° and 100°, and in every case save one the observed surface wind is backed with regard to the gradient wind.

In general it may be said that in this class the pressure is high to the North or North-East and relatively low to the South. It has long been held that on the polar side of a depression the isobars in the higher layers no longer form closed curves; in

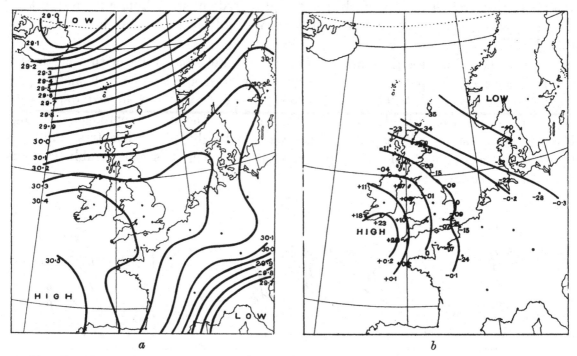

a   b

Fig. 37. Feb. 2nd, 1908, 6 p.m. *a*. Surface isobars. *b*. Computed isobars at 3 kilometres, showing relative differences of pressure from that over Portland Bill.

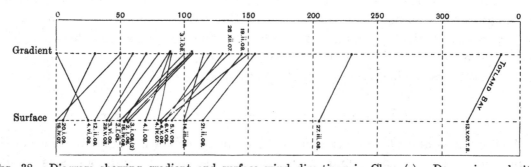

Fig. 38. Diagram showing gradient and surface wind directions in Class (*c*). Decreasing velocity.

the upper layers the depression is a southward extension of the polar low pressure system; the slackening of the wind velocity in the upper layers on the polar side of a cyclone quite corroborates this view.

With regard to the wind direction in this class it has been pointed out above that the surface wind is nearly always backed with regard to the gradient wind. In some cases the gradient direction is never reached; but it is reached or the nearest approach to it is reached a little above the point of maximum velocity. At higher levels there is sometimes a backing towards a Northerly or North-Easterly point, and sometimes the wind remains Easterly up to the highest point reached. In a few cases there is considerable veering in the upper layers; for instance on May 16th, 1909. Examples of an Easterly wind being found at great heights occur on May 6th and 7th, 1909, on Aug. 5th, 1909, and on March 3rd, 1910. On May 6th and 7th, 1909, the direction remained from East to South-East up to the level of the Stratosphere; on Aug. 5th, 1909, it remained between North and East up to 8 kilometres, when there was a discontinuity, the wind above being South-East up to 12 kilometres, when all real wind seems to have ceased. On March 3rd, 1910, the wind after being South-East at 3 kilometres slowly backed to North at 8 kilometres and remained between North and North-East to 11 kilometres when the peculiar wind conditions of the stratosphere were reached.

Class (d). Reversals or great changes of direction in the Upper Layers.

Cases of reversal of wind direction are very fairly common, but they are probably not so common as the number of cases recorded in these observations would lead one to expect; for very often when these conditions were suspected a balloon was sent up when it would not have been otherwise.

The explanation of the reversal is not always easy to determine and probably different causes are at work in different instances. In some cases when the wind velocity is small, as for instance on April 9th, 1907, the change through 360° may be explained by some local eddy, or even possibly by the spiral motion of the balloon during its ascent through a calm atmosphere. In some cases differences of surface temperatures may occasion a reversal of the pressure gradient in the upper air. The cases are considered in detail below.

On Jan. 25th, 1907, a westerly wind was found at a height of one kilometre, the surface wind being from 115°. On this occasion there was a ridge of high pressure over the station, between a deep depression over Scandinavia and a shallower one over Southern Spain. Temperatures were low over the Continent and relatively high over our Western coasts. The warm westerly current due to the Northern depression was evidently rising over the cold current due to the Southern depression. As will be found in further examples this rising of a moisture laden current results in precipitation, and in this example snow showers fell in most places in our islands.

On March 30th, 1907, a surface wind from 250° veered to 55° at a little over one kilometre. There are no indications on the weather map of gradients for North-Easterly winds; this case may possibly be due to the outflow of air from above the depression situated in the neighbourhood of Iceland, and as such should be classed with Class (e), but it is not a typical case. On this occasion there was no rain over these islands except a little in the extreme South-West of Ireland.

On April 1st, 1907, the surface wind was South and it was South again at 4 kilometres, but the intermediate wind was about 115°; the velocity was 3 to

4 metres per second but it decreased to practically a calm at 3 kilometres when there was a sudden change of direction from 155° to 220°; during this day and the next a Southerly wind was replacing an Easterly wind, the Southerly wind spreading from the West; it would seem from this ascent that the Easterly wind still persisted for a time in the middle layers when it had been replaced by the Southerly wind at the surface and also at a considerable height.

On April 9th, 1907, with extremely small wind velocity there was a backing of the wind completely round the compass between the surface and 1·5 kilometre; there was a centre of a cyclone almost over the station, and the temperatures over this country were extremely uniform.

On April 15th, 1907, at 10.50 a.m. it would seem that the shallow depression that caused the surface wind from 35° failed to make its influence felt above one kilometre, and that the wind above was influenced by the low pressure system to the North-West; the wind in this case was very light in the upper layers. It seems probable that the wind system on the South-East of the Atlantic low was bringing warm air from the South, and this current would rise over the North-Easterly current which came from colder regions; in this case one would expect the warm and moist current from the Atlantic rising above the cold current to produce rain, and this was the case, rain having fallen over Southern England and over France. The ascent in the evening of the same day showed a somewhat similar veering of the wind, though at the highest point observed, 1·5 kilometre, it seemed to have backed to a more easterly point.

On May 14th, 1907, there was a surface wind of 30° which gradually veered to 115° at a height of a little less than one kilometre; at the same time the velocity fell nearly to zero; above this the wind direction changed to 325° and then backed to 250° at 2 kilometres with some increase in velocity. The surface wind was probably due to the shallow low whose centre was over Brest, the upper wind being due to the depression over the North Sea. The temperature distribution was very irregular but it seems likely that in this case, as usual, there was a warm westerly current flowing over a colder easterly one; we find also that there was a belt of rain from Spain to the North of England with thunderstorms in places. A map of the isobars at 2 kilometres, Fig. 39, shows that at this height the pressures in the neighbourhood of the North Sea were slightly lower than those near Brest.

On May 17th, 1907, a light surface wind of 305° backed to 240° at 2 kilometres; in this case the upper current certainly came from a warmer region. There was rain on our East coasts.

On May 21st, 1907, a surface wind of 135° backed to 85° at 0·7 kilometre, and then veered to more than 250° at 2 kilometres, after which it backed to 215° with great increase in velocity; the explanation of this curious wind distribution is probably due to the low pressure system situated over Brest being obliterated at the height of about one kilometre; above this height the wind seems to be influenced by the larger depression whose centre was between Scotland and Norway. A map of the isobars at 3 kilometres, Fig. 40, distinctly shows that surface temperatures would

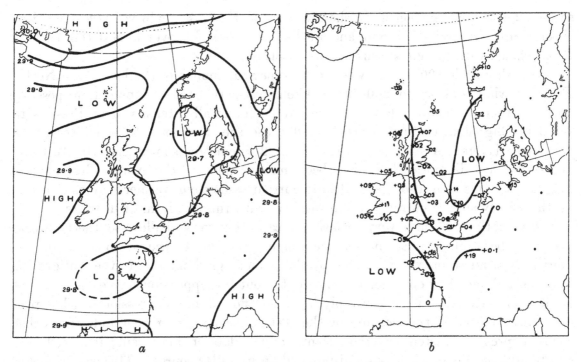

FIG. 39. May 14th, 1907, 6 p.m. *a.* Surface isobars. *b.* Computed isobars at 2 kilometres showing relative differences of pressure from that over Portland Bill.

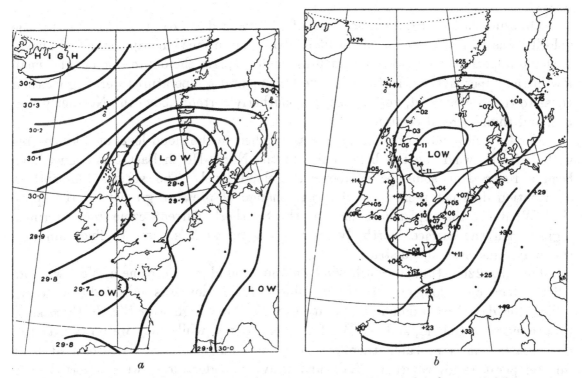

FIG. 40. May 21st, 1907, 6 p.m. *a.* Surface isobars. *b.* Computed isobars at 3 kilometres showing relative differences of pressure from that over Portland Bill.

account for the obliteration of the shallow low. In this case again we have conditions theoretically favourable for rain, and we find that rain fell over almost the whole of these islands with thunderstorms in several places.

On May 25th, 1907, there was a case somewhat similar to the last one, a North-Easterly wind veering to 200° at 2 kilometres; the same explanation possibly applies; there is again a low pressure system of small intensity over Brest with relatively high temperatures which would cause the depression to be obliterated in the upper layers. A map of the isobars at 2 kilometres shows that the Brittany depression has disappeared, or has been moved more to the North-West at this height. In this case also the Southerly current came from warmer regions than the north-easterly surface current; there was rain with thunder in most districts.

May 27th, 1907, is a very remarkable case; the very shallow southerly breeze may have been a local phenomenon, the remains of a day breeze on to the land; the Northerly wind seems to be caused by the general gradient due to the Anticyclone over Iceland and the depressions to the South; but the upper South-Westerly current is difficult to explain; it may possibly be the outflow from a depression which spread over this country on the morning of the 30th, the map giving some indications that such a depression existed over the Atlantic on the date of this ascent; it moved very slowly after reaching our coasts and pursued an irregular course. This ascent should be compared with those in Class (e) and might perhaps have been included in that class. As in other cases of inversions there was rain with thunderstorms in places.

On June 29th, 1907, a surface wind of 340° backed to 175° at 2·5 kilometres; the velocity was 12 metres per second a little above the surface but rapidly fell off above; it would appear that surface temperatures would explain the obliteration of the small low in the upper layers. In this case the Continental temperatures were higher than those to the North-West, and we see the Southerly current above the Northerly one: rain and thunder resulted from these conditions.

On July 1st, 1907, a ballon sonde was watched to the cloud level at 2 kilometres with a North wind; as the balloon fell to the North-North-East there must have been a reversal at some point above the cloud level. The surface temperature distribution does not seem to explain the change of pressure gradient in the upper air. Whatever the explanation may be the Southerly current came from a warmer region than the surface Northerly current; again in this case we find rain and thunderstorms over the district.

On July 22nd, 1907, a South wind on the ground level became a West wind at a little above one kilometre. In the morning a shallow low was indicated over Kent; during the day the maximum temperatures were fairly high over the North coast of France, parts of Ireland, and the West of this country, while they were low over the East coasts; it is probable therefore that the change of wind is associated with this temperature distribution. Rain and heavy thunderstorms were reported over the country.

The case of Aug. 5th, 1907, is further discussed under the heading of " Changes occurring during the day." The remarkable backing of the wind at 2.24 p.m. is no doubt due to the pressure distribution; a V-shaped depression was passing away to the East and influenced the surface wind, while the upper current was probably due to the low pressure system over the Hebrides. Temperatures over the Continent being much higher than those to the West we find the Southerly wind above the North-Westerly one, and as usual we have rain and thunderstorms over the district.

On Aug. 23rd, 1907, the surface wind of 155° differed by 175° from the gradient direction; it is just possible that in this case a mistake had been made in the setting of the theodolite, though reference to the original note book does not support this idea; the falling off of the wind above 2 kilometres and the sudden backing at 3·5 kilometres seems to point to an inversion at a greater height; there were intermediate clouds from West-North-West moving very slowly, with an upper cloud sheet moving more quickly.

On Aug. 28th, 1907, at 11.38 a.m. a wind of 115° at the surface suddenly veered to 235° at 2·5 kilometres. In the evening the surface wind which had become very light came from 165° and backed through East and North to 225° at 1·5 kilometre, remaining very steady above. The temperatures on the whole were higher to the South, but the case is not very clear. The upper wind is due no doubt to the low that was passing up our western coasts. In the morning observations the upper wind is strongly curved outwards from the surface isobars. There was rain in many places and lightning was reported in the Channel at night.

On Oct. 14th, 1907, there was a very remarkable case of inversion; at the surface the wind was from 340°, and it rapidly rose to more than 10 metres per second at a height of about half a kilometre; at a height of 0·8 kilometre it had fallen to a calm, but rapidly rose again to more than 10 metres per second at one kilometre, with sudden veering through 200° to direction 190°, the gradient direction being about 215°. A squall had just passed by with heavy rain, and the ascent was made because scud was seen to be moving from opposite directions in different layers. The 6 p.m. map shows a small centre of low pressure over the South-Eastern counties: this disturbance had evidently just passed the station when the ascent was made.

On March 3rd, 1908, the wind veered considerably in the first half kilometre, going from 165° to 235°; the direction then slowly backed to 215° at 2·5 kilometres when there was a sudden veer to 295° followed by a slow backing to 225° at 3·5 kilometres. The sudden change is without any adequate explanation; the centre of a shallow low pressure system lay over the North of England on the morning of this day.

On March 12th, 1908, a surface wind from 215° veered to the North, the change being nearly complete at 0·9 kilometre. In this case the lower temperatures on the Continent account for the obliteration of the shallow low over our East coasts and intensify that over Europe in the upper air. On March 13th the surface temperatures were again favourable to gradients for North-Westerly winds in the upper air.

In both cases the warm winds from the Atlantic seem to rise over the colder current from the Continent, and the result was that there was rain or snow in many places.

On March 21st, 1908, the surface wind of 135° veered to 200° at 2 kilometres; the upper wind which blew outwards from the surface isobars was no doubt due to the advancing low off the West coasts; the warmer air from the Atlantic is here again seen to be flowing over the colder air from the Continent; there were showers of rain and hail over the western districts of our islands.

On April 9th, 1908, the surface wind was from 135°; it was very light and proved to be a shallow current, only half a kilometre thick; the upper wind was due to the general gradient. The temperatures were rather irregular but on the whole were higher in the West. There was rain over most parts of the British Islands.

On June 2nd, 1908, the surface wind of 145° veered to 240° in the first kilometre; the gradients are slight and irregular, and the temperatures over the Continent being relatively high the pressure in the upper layers would be lowest to the North-West; this would occasion South-Westerly winds over the station in the upper air. June 2nd is an exception to the general rule that the upper current comes from the region of warmer temperatures; the upper wind was from a little West of South while the surface wind came from the South-East where the temperatures were the highest; it is possible however that the surface air really was drawn from the region of lower temperatures over the North Sea, and that the current only became a South-Easterly wind near the station. On this day rain and thunderstorms were recorded.

On July 27th, 1908, there was a curious case of a light Westerly wind veering to about 200°; the change was complete at 1·5 kilometre, above which the direction remained extremely steady. In the upper layers this case corresponds to Class (b), the wind velocity increasing to 25 metres per second at 8 kilometres. The high temperatures over the Continent and the relatively low ones near the low to the North-West would increase the pressure gradient in the upper layers, and would also result in the Southerly wind flowing over the Westerly; rain occurred but in this case thunderstorms were not reported.

The two ascents on the morning of Sept. 30th, 1908, show a remarkable phenomenon; after a slight veering up to 1 kilometre there is a great backing of the wind accompanied by a complete falling off of velocity; above the calm layer there was a discontinuity and the direction above is not very different from that at the surface. At still greater heights the wind behaves like a Southerly wind of Class (b), the velocity increasing up to the greatest heights reached. It would appear that the lower Southerly wind is related to the high pressure system over the Continent, while the upper Southerly wind seems to belong to the system of low pressure over Iceland.

On November 3rd, 1908, with an Easterly surface current of small velocity there was a steady backing of the wind direction from 130° at half a kilometre to 300° at

4·5 kilometres. This was possibly the commencement of the Westerly wind that was found in the upper layers on Nov. 7th, 8th, and 9th. The velocities were small but were increasing slowly above 4 kilometres.

On Nov. 6th and 7th, 1908, the conditions were very remarkable; a strong Easterly wind on the surface became a North-Westerly wind at 4 kilometres, there being a layer of calm air between 3 and 4 kilometres on Nov. 6th; on the 7th the wind changed at a lower level, and the upper wind was from the West. On both days there were low temperatures over the Continent and high temperatures on the West coasts which to a certain extent would account for the reversal of the wind in the upper layers. A study of the weather maps for these days shows an Easterly or North-Easterly wind over Western Europe and a Westerly or North-Westerly wind over Eastern Europe; it seems as though the upper Westerly wind descended at the region of highest pressure, part of it going on to make the Westerly wind of Eastern Europe, part of it flowing back to make the Easterly wind of Western Europe. On the 6th there was some rain on the East coasts of England and Scotland, and on the 7th over South-Western France; on both days there were clear skies over Germany where presumably there was a current of descending air.

On the following day, Nov. 8th, somewhat similar conditions were maintained; the strong Easterly wind at the surface fell off from 25 metres per second at 1 kilometre to less than 5 metres per second at 2·5 kilometres; at the same time the direction which was 90° at 2·5 kilometres backed steadily to 300° at 4·5 kilometres. Neither surface pressures nor temperatures would lead one to expect an upper Westerly current.

These three cases are very instructive as they show the wind conditions that occurred during the passage of a well marked but not very deep depression to the South of this country. On the 7th there was a depression over the Western Mediterranean; this moved eastward during the day and decreased much in intensity, but it was followed by another which came from the Atlantic and was off the North-West coast of Spain by the evening of the 7th; by the evening of the 8th its centre lay over the Southern coast of France. On all three days there is evidence of the upper wind descending and dividing into two parts, one going onward, the other flowing back under its own course in the upper air.

On June 1st, 1909, we have a case of a Northerly surface wind changing to a Southerly wind at a height of 2 kilometres; in this case it appears that the Northerly wind was forcing its way under a warm Southerly current and producing rain as shown on the weather chart for that date. The upper wind seems to have belonged to the system of the Continental Anticyclone; the lower to the advancing Atlantic high pressure system which caused a flow of cold air from the Arctic regions.

In practically all the cases of this class we find that the upper wind is flowing from the warmer regions, and the lower one from the colder; so that the upper wind is, if not absolutely at any rate potentially, warmer than the lower. In almost every case we find rain occurring as a result, as would be expected if a warm current of air rose over a cold one. In most cases too we find thunderstorms occurring. In fact it seems possible that a reversal or a great change of wind direction in the upper layers may be necessary for the production of lightning; it is hard to suppose that a thunderstorm would go on for more than a short time unless electricity were being conveyed to the region of the thunderstorm from elsewhere; if one current were at a different electrical potential from the other the supply might be maintained.

In connection with the subject of reversals it must be noticed that when there is a discontinuity or sudden change of wind direction the velocity always falls off to a very small value; even with a sudden veering or backing of the wind this is the case. This must be borne in mind in what has been stated above about one current flowing over another. The old idea of one current flowing close over another, and producing waves, as wind produces waves over the sea, receives no support either from theoretical considerations or from actual observations. It is obvious that the wind at any height must be the result of the pressure gradient at that height, and it does not seem possible that the pressure gradient should change abruptly between adjoining layers of the atmosphere. Calm layers mark the boundary between comparatively sudden changes of wind direction.

Class (e).   Upper wind blowing outward from centres of low pressure.

We now come to the most remarkable class of the series. It is frequently found that the wind in the upper layers blows outward from a centre of low pressure, and is sometimes nearly at right angles to the surface isobars. More for the sake of convenience of reference than anything else the class has been divided into two subclasses; (1) upper wind from between West and North; (2) upper wind from between South and West.

It will perhaps be best to take each case separately as in this way the peculiar features of the class can best be appreciated.

1. Upper wind from between West and North.

On Feb. 8th, 1907, a Southerly wind at the surface changed suddenly at 0·7 kilometre to one from 300°; there was a depression whose centre was to the South-East of Iceland and the upper wind blew out from this nearly at right angles to the surface isobars; this depression rapidly increased in intensity during the following days, but it did not change its position materially till Feb. 11th.

On April 18th, 1907, the central part of an anticyclone was over the station. Far out to the South-West of Iceland there was apparently a low pressure system; the gradient velocity was 5 metres per second and the direction 300°; the surface velocity was a little less than the gradient and the direction was 20°; at 5 kilometres the velocity had increased to about 17 metres per second and the direction was 325°; a map of the isobars at 4 kilometres, computed from the surface temperatures, shows a gradient for North-Westerly winds but the gradient is not steep; in this case the direction of the upper wind was in agreement with the surface isobars. By the following evening the anticyclone had moved South-West and the Icelandic depression had become more marked; the wind velocity was small and the direction South at the station, but at a height of one kilometre it had veered to the North; this change is quite inexplicable by surface pressures or temperatures; a map has been drawn to show the isobars at 3 kilometres deduced from the surface pressures and temperatures only and it shows no gradient for Northerly winds. By the 20th the Icelandic depression was very marked with a much steeper pressure gradient over this country. The gradient direction was 225°, the surface direction 190°; at 4 kilometres the direction observed was 345°; the velocity was decreasing at the highest point observed. On all three days the wind in the upper layers was blowing more or less away from the advancing low pressure system. This depression moved slowly onwards and by the 24th it was in the neighbourhood of Shetland. On all three days the persistent Northerly current cannot be accounted for by surface pressure or temperature distribution.

On Sept. 4th, 1907, there was a sudden change of wind from South to West just above 1 kilometre; the lower wind was probably due to the high pressure over Germany, the upper to the low over the Atlantic; the temperatures are rather irregular but air drawn from the South-West would probably be at a higher temperature than that drawn from central Europe. The upper wind was blowing directly out from the low pressure system, at right angles to the surface isobars. As usual in cases of great change of wind direction rain fell, and in this case in nearly all parts of these islands.

On Sept. 23rd, 1907, with a surface and gradient wind from 110° there was a fairly regular veering to 300° at 4·5 kilometres; the surface isobars gave no indication of the North-Westerly wind which seemed to be blowing out from the centre of the low pressure system over Iceland. This depression moved down to our Western coasts by the morning of the 25th, and thence reached the Bay of Biscay on the 27th. Above 4·5 kilometres the direction was very irregular, and at all heights the velocity was extremely small, never exceeding 5 metres per second. There was slight rain on our far Western coasts.

On Jan. 11th, 1908, the wind direction was 170° on the surface, the gradient direction being 240°, which direction was reached at 1 kilometre; above this the wind veered to 90° at 2 kilometres and (after some backing) to 100° at 4·5 kilometres; it then backed to 25° at 6 kilometres; the velocity was small but was increasing at the greatest heights. There were high temperatures to the North and low to the South but temperature differences on the surface would not account for the reversal at 2 kilometres. The Icelandic depression from which the North wind was perhaps an outflow, moved slowly to the South-East towards these islands.

On Feb. 14th, 1908, a West wind fairly constant in velocity and direction up to 4 kilometres, veered above that height to 330° with rapid increase of velocity; as usual in cases of this class there was a low pressure system near Iceland, and it moved rapidly across to Scandinavia in the course of the following 48 hours.

On March 5th, 1908, the wind veered from 225° at the surface to 275° at 3·5 kilometres; it was veering rapidly at the last point of observation and increasing in velocity; the direction at the highest point was strongly outcurved from the surface isobars. A deep depression to the South of Iceland moved rapidly towards these islands and by the following morning had its centre over the Irish sea. The lower wind was from about 250°; this backed and fell to a calm at 1·5 kilometres; it was probably part of the system of low pressure over Holland which was decreasing in intensity and passing away during the course of the day. The ascent in the evening showed that this lower wind had died out altogether and the lowest strata were now influenced by the approaching system to the West.

On March 24th, 1908, there was a very striking case of a South wind on the surface changing to a North wind at 3 kilometres. There was no indication from the surface isobars or surface temperature distribution of such a complete change of direction. There was a shallow depression to the West which by the following morning had moved to the South of England. There was also a low pressure system near Iceland.

On March 27th with a very similar pressure distribution a similar change of wind direction took place, a surface wind of 200° veering to 335° at 2·5 kilometres.

On March 28th, 1908, the upper wind was blowing nearly at right angles to the surface isobars and away from the low pressure area over Iceland; this system moved to the South-East and was over the Shetlands on the morning of the 31st.

On April 1st, 1908, there was a great increase in velocity with height, and the direction veered from 295° to 320°; some increase in velocity can be explained by surface temperatures, but so great an increase in velocity would not be expected, and the wind in the upper layers was blowing outwards across the surface isobars. In this case the low to the South-West of Iceland (from which the upper current appears to be an outflow) moved towards this country being over the Shetlands on April 3rd; it then passed away across the North Sea.

On April 29th, 1908, there was a very marked veering of the wind from 215° at the surface to 300° at 4·5 kilometres above which it remained relatively steady to 6 kilometres; above 4 kilometres there was a rapid increase of velocity, which rose from 8 metres per second to over 20 metres per second at 6 kilometres. There was a depression over the Atlantic which moved nearer to these islands, and was just off the North of Ireland by the following morning, after which it moved out over the Atlantic again.

On June 10th, 1908, a West wind at the surface backed to 220° at 1·5 kilometre, and then steadily veered to 320° at 6·5 kilometres. There was the usual depression in the neighbourhood of Iceland, which subsequently moved across to Scandinavia.

On June 22nd, 1908, a light Southerly wind on the surface became a Northerly wind at 1·5 kilometres above which there was a slow veering till at 12 kilometres the direction was entirely reversed from the surface wind. There was a low pressure system over Iceland which moved slowly towards Scandinavia during the succeeding days.

On July 28th and 29th, 1908 (see diagrams under class (f), the Stratosphere), the wind was Northerly up to very great heights and increased in velocity slowly but steadily. There was a rather deep depression over Iceland which moved Eastward and by the 31st was over Scandinavia. On the evening of the 30th when the centre of the depression was due North of this country the wind was North-Westerly and there was not such a marked increase in velocity in the upper layers as on the previous days. On the ascent in the morning at 8 h. 6 m. the wind increased and veered from 270° on the surface to North at 2·5 kilometres; at 0·36 p.m. there was a little veering but the increase was less marked; this balloon was only watched up to 4 kilometres. By the evening of the 31st however with the wind remaining in the same direction in the upper layers there was again a great increase in velocity, 20 metres per second being reached at 4 kilometres and 35 metres per second at 11·5 kilometres, the direction remaining steadily in the North-West. On Aug. 1st both in the morning and in the afternoon the Northerly wind steadily increased to values greatly exceeding the gradient value at 3 to 4 kilometres, and though by the evening of the 2nd what little wind there was at the station came from a Southerly point, yet at half a kilometre it had gone round to the North, and at 4 kilometres it was blowing at the rate of 15 metres a second or more. It is significant that by the evening of the 31st there are signs of another disturbance off the West of Iceland which advanced and passed in an Easterly direction along the Arctic circle on the night of the 2nd.

On Nov. 16th, 1908, the wind direction went completely round the compass between the surface and a height of 3 kilometres; the lower Northerly wind was part of the circulation due to a high pressure system which was approaching these islands from the Atlantic; the intermediate Southerly wind, which was very light, was perhaps due to the high pressure system over the Continent combined with the low over Iceland; the upper Northerly wind, which was increasing in velocity at the highest point of observation, seems to be an outflow from this depression, which moved towards Scandinavia during the following 24 hours.

On Feb. 7th, 1909, there is a case of a South-Easterly surface wind changing to a North-Westerly wind at 2·5 kilometres. The temperatures over the Continent were low and those over the West Coasts were relatively high, but the surface temperature differences alone are not sufficient to account for the change of direction at 2·5 kilometres. The weather map for this day shows a phenomenon somewhat similar to that mentioned under Nov. 7th and 8th, 1908; namely that the upper wind seems to descend over Germany, part flowing on as a North-Westerly wind, part flowing back again as the South-Easterly surface wind. There was a low pressure system over Iceland which advanced and whose centre was over this country on the 10th. There was rain over Ireland on this day.

On Feb. 17th, 1909, the wind went right round the compass; it was 290° at ground level and remained very steady up to 1 kilometre when it backed rapidly, being about 50° at 2 kilometres; above this the backing was slower; the direction was 295° at 6 kilometres. On the following day the surface wind was 115° and remained steady but with decreasing velocity up to 3·5 kilometres; above this there was a sudden change, a discontinuity; at first the wind was East of North but it

7—2

finally backed to West of North and was 335° at 8 kilometres.   A map of the isobars on Feb. 18th for 5 kilometres, deduced from surface pressures and temperatures shows very little gradient ; a similar map for 8 kilometres shows some high pressures to the West and low pressures to the East which would give a gradient for Northerly winds ; but it may be doubted whether it is legitimate to consider the isobars at so great a height as 8 kilometres in this way.   On both days there was a depression to the West of Iceland which was increasing somewhat in intensity.

On March 5th, 1909, there was a surface wind of 160° that veered to 270° at 4 kilometres with considerable increase in velocity.   There was a depression off our South-West coasts which advanced over the country in the course of the following days ; the upper wind was blowing directly outward from the centre of the depression and nearly at right angles to the surface isobars.

On May 2nd, 1909, a surface wind of 250° veered to North with great increase in velocity ; there was an extensive low over the Atlantic.

There are two cases in this class that resemble one another remarkably, namely March 24th and June 22nd, 1908.   In both cases there was an anticyclone over North-West Russia, and a low to the West of Iceland ; in both cases there was a V-shaped depression to the West of the station ; in both cases the wind veered remarkably ; on March 24th the veer was from 230° to 10°, and occurred between 2·5 and 3 kilometres ; on June 22nd it was from 240° to 25° and took place in the first 1·5 kilometres.

2.   Upper wind from between South and West.

There are a few cases in which the Southerly upper current seems to be the outflow from a low pressure system situated to the South of our area.

On July 24th, 1907, the wind backed from 140° at the surface to 10° at just under 1 kilometre, and then veered to 220° ; the lower current was probably due to the depression over the South-West of France.   The pressures were rather irregular but it may be that the South-Westerly wind represents an outflow from the belt of low pressure to the South.   We find on this day that there were thunderstorms in the South of France and rain over Ireland and part of England.

On May 2nd, 1908, the gradients were irregular and slight ; the upper wind was blowing directly away from the low pressure system over the Atlantic to the West of Ireland ; as we saw in the cases of inversion under class (d) the Easterly wind must have come from colder districts than the Westerly wind that flowed over it ; there were thunderstorms at night over the West of England.

On June 3rd, 1908, the pressure was low over Spain and high over the Baltic. In both morning and evening ascents on this day the wind veered from North-East at the surface to about South at 5 kilometres ; the surface wind in the evening agreed closely with the gradient ; the upper current may very likely have been an outflow from the low pressure system to the South.

On May 16th, 1909, the wind veered from 50° at the surface to 190° at a little over 2 kilometres ; the upper wind was blowing directly outwards from the low pressure system to the South, and it is worthy of note that this depression moved North-East across the North Sea on the subsequent day. Rain occurred in many parts of the country.

On June 3rd, 1909, the wind at the surface was 35° and it veered to 155° at 3·5 kilometres; in this case also there was a depression to the South-West, and this took a course very similar to that of the depression noticed on May 16th; it crossed the North Sea on the night of June 4th.

CHAPTER VI

CHANGES OF THE WIND DURING THE DAY AND DURING CONSECUTIVE DAYS

March 28th to April 9th, 1907.

ON March 28th there was a high pressure system to the East with its centre over Holland and a depression to the Westward of Iceland; on the evening of this day, above a surface irregularity, the wind was East with increasing velocity, a current evidently due to the high pressure system. By the following evening the latter had moved Eastward and the wind veered from South-East at the surface to South at 1 kilometre, the type of vertical wind distribution being midway between classes (a) and (b), and being due to the low pressure system which was extending from Iceland. By the evening of the 30th a high pressure system of small intensity had appeared over South-Western France and the South-Westerly current due to the Icelandic low veered rapidly with height to a North-Easterly wind at 1·5 kilometres, the upper wind being apparently due to the French high pressure system. By the evening of April 1st the depression over Iceland had moved away to the North-East, and another depression had appeared to the West of Iceland. A Southerly current was now replacing the Easterly wind both at the surface and at 3 kilometres, but the Easterly wind persisted in the intermediate layer, till on the following day, the Southerly current had entirely replaced it; the type of vertical distribution now belonged to class (b) with rapid increase in velocity in the upper layers; the low pressure system had approached these islands from the North-West. On April 3rd there was no ascent. By April 4th the depression had moved to the South-East of France, but another low was approaching these islands from the Atlantic; on this day the vertical distribution was of class (c) with wind backing from East at the surface to North at 3 kilometres; the Northerly wind was of extremely small velocity pointing to a probable change of direction at a still higher level. By the evening of the 5th, there was a South-Westerly current up to 4 kilometres, which seems to have been related to the low pressure system now centred off the North-West coast of Ireland. By the evening of the 6th the pressure distribution had not materially changed and we find that the vertical wind distribution was much the same as on the evening of the 5th. On the 7th there was no observation. By the evening of the 8th there was a large shallow disturbance over the greater part of these islands, the station being just to the South of it; the wind was Westerly and did not change much either in velocity or direction up to 3 kilometres. By the evening of the 9th the centre of the depression was practically over the station; the wind was very light but its direction changed rapidly with height; at the surface it was Southerly but between 0·7 and 1·6 kilometre it backed completely round the compass, becoming Southerly again at the latter height.

Taking the series as a whole the upper wind was obviously generally related to the coming type of pressure distribution, while the lower wind appears in most cases to be the last effect of the pressure type that was passing away. The upper wind was generally parallel to the surface isobars, except on April 5th when the wind at the highest point observed was blowing outward at an angle of 50° to the gradient wind.

April 15th to April 20th, 1907.

On the morning of April 15th the surface wind was North-East and was due to the low pressure over the South of France; the surface wind rapidly died away and at 1 kilometre it was replaced by a South wind due to the low pressure off the West coast of Ireland. In the evening the surface wind had backed to nearly North, the French depression having moved meanwhile further to the East; the wind fell off to almost a calm at 1 kilometre, and veered through 140°; but the last two observations showed that it backed again to about 70°. By the evening of the 16th the small low over South-Eastern France had considerably increased in size; the wind was North at the surface veering to North-East at 1 kilometre, the velocity decreasing with height. By the evening of the 17th the depression to the South-East was represented by a large area of low pressure stretching from the South of Sweden to Tunis, while there was a large area of high pressure over the Atlantic; the isobars over this country were parallel, but there were some irregularities; the wind was Northerly and except for some irregularity near the surface it remained steady in velocity and direction up to 2 kilometres; the typical Easterly type of vertical distribution which had occurred on the previous day was now at an end. By the 18th the high pressure from the Atlantic had moved over these islands; it proved to be a narrow ridge and beyond it to the South-West of Iceland another depression was making its appearance; the surface wind was 30° backing to 315° at 1·8 kilometre, the velocity being extremely small; at 2 kilometres there was a sudden veer to 355°, and above this the velocity rapidly increased with height; the direction backed to 330° at 4·5 kilometres; the upper wind seems to have been the outflow from the low pressure system to the North-West. By the evening of the 19th the centre of the high pressure system was to the South-West of the station, over the Channel, and there was a light Southerly breeze due to the anticyclonic circulation; the wind rapidly veered and was North at 1 kilometre; here again we have an apparent outflow from the Icelandic depression; there was no increase in velocity up to 2·5 kilometres but above this there were signs that an increase was beginning. On the 20th the high pressure system had moved a little more to the South, and there was now a long belt of high pressure from Prussia to the North-West coast of Spain; the Icelandic depression was much deeper and was advancing; on the surface the Southerly wind was maintained and had increased since the previous day; after the first half kilometre it fell off, and the direction veered steadily to 355° at 4·2 kilometres; an upper current blowing directly away from the Icelandic depression was thus maintained for three successive days.   No further observations were made during the following days, but the depression moved slowly Eastward and reached Norway by about the 23rd.

Jan. 2nd to Jan. 4th, 1908.

There were six ascents on these days; during the whole time there were low pressures to the South causing Easterly winds of some strength on the surface; in all the cases the wind decreased with height, and veered about 30° in the first kilometre; there was very little change in the general pressure distribution, and very little change in the vertical wind distribution during these days.   The Easterly winds probably persisted up to considerable heights, for though the observations were not carried beyond a few kilometres, two of the balloons of Jan. 3rd were found over 60 kilometres to the West.   It is noticeable that no low pressure areas are shown on the Weather charts to the West or North-West.

May 29th to June 5th, 1908.

On May 29th, at 8 a.m. a large anticyclone covered the whole of the North Sea and the Southern half of Scandinavia; the vertical wind distribution was typically Easterly, class (c).   By the evening of the following day the anticyclone had moved somewhat to the North-East, and the low pressure over Spain had extended towards our area; the vertical wind distribution was of class (a) with some veering of the wind, which was Easterly at the surface, and little increase in velocity up to 7 kilometres.   No definite low pressure is shown on the Weather chart to the West or North-West.   By June 1st, however, a low pressure system was approaching these islands from the West, and the high pressure was now over the Arctic circle between Iceland and Norway; a surface wind from the East veered to a Southerly wind in the first kilometre, and above this the distribution was of class (b), the wind increasing in velocity but remaining steady in direction.   By the evening of the 2nd, the Northern anticyclone had moved to the North of Scandinavia and the low pressure system had decreased in intensity, while a low over Spain had extended Northwards almost to the mouth of the Channel, in fact the pressure distribution was very similar to what it had been on May 30th, except for a small low over the North of Ireland; the vertical wind distribution has been included in class (d) as there was a change of direction from 135° to 240° in the first half kilometre, but above this the wind was fairly steady both in velocity and direction (South) up to 6 kilometres.   On June 3rd, there was a high pressure system over the South of the Baltic region, and low pressures over Spain; there were also small thunderstorm depressions over various parts of Western Europe; the wind velocity remained small up to 3 kilometres, but increased at greater heights; the direction veered from 45° at the surface to 180° at 4 kilometres, and the ascent has been included in class (e 2), the upper current being apparently an outflow from the low pressure areas to the South, flowing towards the high pressure in the North-East. By the evening of June 4th the pressure distribution had entirely changed; a high pressure system lay over the Atlantic off the Irish coast, and pressures were relatively low to the South and to the North-East; we find a vertical wind distribution of class (c), a surface wind of 25° backing to 300° at 2 kilometres and the velocity falling off rapidly. Finally on June 5th we have a very similar pressure distribution, but the gradients are steeper and we find a vertical distribution of wind of class (b), a North-Westerly wind with velocity increasing with height.

July 27th to Aug. 2nd, 1908.

This series of ascents was made in connection with the series arranged by the International Commission for Scientific Aeronautics. On July 27th there was a Westerly wind on the surface which fell off to a calm a little above 1 kilometre; this surface wind seems to have been due to the anticyclone to the South-West; above this the wind veered through nearly 300°, and remained steady at about 215°, but the velocity increased considerably with height; in fact above the reversal the wind behaves as in class (b); it is evidently related to the low pressure system between Scotland and Iceland. By the following day the conditions had completely changed; the high pressure system had increased, and reached our Western coasts, while the low pressure system had receded to the neighbourhood of Iceland; the upper wind had entirely changed and was now blowing with great velocity from the North, though it was very light on the surface. The two first ascents of the 29th are included in class (a), but the observations were only carried to 4 and 5 kilometres, and at greater heights it is probable that the strong Northerly wind of the preceding day and of the following days would have been met with, as it was in the evening ascent of the same day. During the following days the pressure distribution remained much the same; the anticyclone was found with its centre just to the West of our Islands, and there were low pressure systems in the neighbourhood of Iceland or passing along the Arctic circle to the East of Iceland; during the whole time the upper wind was North or North-West and increased with height to very large velocities in the upper layers.

Sept. 30th to Oct. 2nd, 1908.

The first two ascents of Sept. 30th show the curious phenomenon of a surface Southerly wind dying out in the first two or three kilometres and being replaced by a South-Westerly wind above; the lower wind seems to be the result of the high pressure system over the Continent and the upper one of the low pressure system near Iceland. The succeeding ascents all belong to class (b), Southerly winds increasing considerably with height. On these days there was a region of high pressure over Central Europe, these islands lying well to the West of it. The upper winds were in accordance with the surface pressure distribution, but in the higher regions they much exceeded the gradient velocity. A comparison between these ascents and those of the end of July and the beginning of August is very instructive.

Nov. 6th to Nov. 8th, 1908.

On Nov. 6th there was a complete reversal at 3 kilometres, the upper wind flowing towards the high pressures to the East, the lower wind being the South-Easterly current due to the high pressure over the North Sea. By the following day the low pressure to the South-West had spread towards these islands and the high pressure area was a comparatively narrow belt stretching across Europe from Asia Minor to Iceland; the Easterly wind had moderated and the upper Westerly current was found at 2·5 kilometres. By the evening of the 8th, the dominant feature of the pressure distribution was the low over the South of France; the Easterly wind had increased greatly in force and extended to a height of quite 3 kilometres; the transition to the Westerly wind was not so sudden and the wind backed instead of veering. In these examples we have an Easterly wind on the surface replaced by a Westerly wind in the upper layers, a state of things that persisted for three days at least; the upper wind was apparently flowing more or less away from the low pressures and directly towards the high pressure area, where it seems to have descended to the surface.

Feb. 12th to Feb. 24th, 1909.

On Feb. 12th there was a high pressure system centred over Scotland; the vertical wind distribution was of an Easterly type, with high velocities near the surface, decreasing with height. By the 13th the high pressure centre had moved to Ireland and the wind distribution was Northerly of class (b). On the 14th the conditions remained the same. On the 15th the wind was rather more Westerly, but in general the conditions both as regards pressure distribution and vertical wind distribution were similar to those on the previous days. By the morning of the 17th a low pressure system which had been over the Baltic region had moved away to the East, and the anticyclone which had been stationed over our Western coasts had moved to Western France and the Bay; at the same time a low pressure system had appeared off the coast of Iceland, and produced an upper wind from 300°; the lower wind was from 290° but this died out at about 1 kilometre, with a veer through nearly 360°. By the 18th the high pressure had moved to Germany, and the surface wind was from about 115°; this died out with little change in direction at about 3 kilometres, above which the wind was more or less Northerly, blowing out from the low pressure near Iceland. By the morning of the 19th, the vertical wind distribution had changed to class (c), a South-Easterly wind decreasing in velocity with height; low pressure areas lay to the North-West and South-West of the British Isles. By the evening of the same day the South-Easterly wind extended with little diminution of velocity after the second kilometre to a height of 6 kilometres. By the morning of the 20th the conditions were much the same, but the wind direction slowly veered to South; similar conditions prevailed on the evening of the 20th and at 2.30 p.m. on the 21st, though on the latter occasion the velocity decreased somewhat in the upper layers.   On the morning of the 22nd, however, with a pressure distribution very similar to that on the previous days the wind backed from 80° on the surface to 40° at 3·5 kilometres, with a decrease of velocity; by the evening of the 22nd the wind backed in a similar manner but the velocity remained constant, the high pressure to the East having meanwhile extended somewhat in a Westerly direction.   On the two following days these islands were included in the Western extension of the Continental anticyclone; on the 23rd the surface wind was 15° and there was not much change either in direction or velocity up to 3·5 kilometres; on the 24th the surface wind was 70° and backed to about 50°; it was fairly constant in velocity up to 5·5 kilometres.

May 4th to May 7th, 1909.

On May 4th there was a large anticyclone with its centre over the South of the Baltic region; the wind was South-Easterly, decreasing in velocity above 2 kilometres.   By the evening the anticyclone had its centre further North and had somewhat decreased in intensity; the surface wind was Easterly and veered to 140° at 1·5 kilometres, backing to about 100° at 3 kilometres; the velocity steadily decreased with height.   On the morning of the 5th the Easterly type was maintained; there was now a low pressure system over the Western Bay which had apparently moved down from Iceland during the preceding day; by 2.30 p.m. the surface wind from 65° veered slowly to 145° at 6·5 kilometres; by 6.40 p.m. the surface wind was 100°, and in the upper layers the distribution was similar to what it had been in the ascent earlier in the afternoon.   On the 6th the pressure distribution remained about the same as on the previous day; the Easterly surface wind decreased in velocity above 2 kilometres.   By noon both velocity and direction were very steady after a veer from 85° to 130° in the first 2 kilometres.   By 6.30 p.m. the wind distribution had not materially changed, but there was the usual increase in velocity just below the level of the stratosphere   On the 7th the anticyclone had its centre over the North Sea, while a low pressure system occupied the Bay; gradients were consequently steeper over this country; the Easterly wind was strong on the surface but decreased steadily in the upper layers, both in the afternoon and in the evening.

Changes during the day.

On May 10th, 1907, there were two ascents, both showing great increase of velocity with height; a depression lay off the west coasts of these islands; its position had not changed materially between 10.22 a.m. and 6.32 p.m., the times of the ascents, though the gradient velocity had fallen from 12 to 9 metres per second. In the morning ascent the gradient velocity was reached at just over a kilometre; the wind increased rapidly to 9 metres per second but did not again increase till 1 kilometre was reached when it increased steadily to 20 metres per second at 2 kilometres.   In the evening ascent the gradient wind was reached in 0·3 kilometre and by 1 kilometre the velocity was 15 metres per second; there was then little increase till 2·5 kilometres when the velocity increased to 20 metres per second. Above half a kilometre the direction was constant and nearly the same in both cases, agreeing closely with the gradient direction. It seems possible in this case that diurnal convection currents are responsible for the changes; the air from different layers being mixed high velocities are found near the ground in the evening, but on the other hand the current in the upper air has been slackened by the mixture of more slowly moving air from the surface layers.

On Aug. 5, 1907, at 2.24 p.m. there being a V-shaped depression just to the East of the station, the surface wind of 55° backed to North at 0·5 kilometre, and after remaining constant for 300 metres rapidly backed to 215° with increasing velocity; there seems little doubt that the lower wind was related to the V-shaped depression, the upper wind to the deeper low which lay off the West coast of Scotland; by 7.19 p.m. the surface wind had gone to 245°; it veered to 325° at 0·9 kilometre and then backed to 205°; it seemed as though the Westerly current due to the Atlantic low was beginning to creep under the wind of the receding V-shaped depression, which, however, was still dominant in the intermediate layers.

Conditions somewhat similar to those of May 10th were found on Aug. 28th, 1907, when the reversal which at 11.38 a.m. was found at a little over 2 kilometres by 6.23 p.m. was complete at 1 kilometre. On March 27th, 1908, too the reversal that at 11.4 a.m. had been found at 2 kilometres appeared to have descended to the surface at 5.40 p.m. but there had been a considerable change in the isobars in the interval.

On April 9th, 1908, by the evening ascent a southerly wind had begun to creep under the northerly one, there having been no sign of it in the morning; this was probably the remains of an on-shore breeze during the day.

On June 3rd, 1908, the velocities were small; the morning ascent at 10.19 a.m. showed a regular veering of the wind from the ground level where the direction was 60° to 3 kilometres where it was 175°; by 6.54 p.m. the veer was from 50° to 125° in little over half a kilometre; the direction then remained fairly constant for about a kilometre when another veer occurred.

On Aug. 5th, 1908, there seemed to be a decrease of wind velocity at the height of about 1 kilometre that became more marked in the evening; it was not associated with any change of wind direction; it might possibly have been the result of an on-shore day breeze further in the direction of the coast, such as made itself felt at the station on the evening ascent of the following day.

There are cases also, Feb. 19th and May 4th, 1909, for example, where the increase often observed above the surface with easterly winds is partially obliterated in the evening ascent. There is thus some evidence from these ascents of what was already known, that the effect of diurnal convection currents is to cause stronger winds on the surface and lighter winds a short distance above as the day advances. The evidence from the pilot balloon ascents is not very striking for clear weather in which balloons can be observed is not weather in which convection currents are strong.

CHAPTER VII

The wind in the Stratosphere

In a few cases the balloon has been watched to great heights, so that its trajectory could be plotted after it had entered the Stratosphere or Isothermal Layer. As is well known the investigation of the upper air by means of balloons carrying instruments has revealed the fact that while the temperature falls more or less regularly with height up to a certain point, yet after this the temperature remains nearly constant up to the greatest heights reached. The temperature in this layer, named the Stratosphere, varies from place to place and from day to day, the variations being very large; the temperature from one day to another may change as much as the change from summer to winter on the surface; the layer is therefore only isothermal in the sense that its vertical temperature distribution is isothermal over a certain place at a certain time.

The number of cases in which balloons have been followed into the stratosphere is small; at Ditcham there have only been eleven cases. To judge by these it appears that just below the beginning of the stratosphere the wind velocity reaches its maximum value, and that when the layer is entered there is a more or less rapid fall in the wind velocity; the direction does not seem to change very much until even greater heights are reached. It will perhaps be best to consider each case separately. From July 28th to July 31st, 1908, four balloons were seen after they had presumably entered the layer, but unfortunately only one of the instruments carried by the balloons was recovered, all the balloons falling in the sea. The pressure distribution for these days was as follows: on the 28th there was an anticyclone to the West of these islands with a pressure of 30·4 inches; over Finland there was another maximum of pressure with a pressure of 30·3 inches; over Iceland was a large and shallow low pressure system. By the 29th the anticyclone had moved rather more to the East the barometer standing at 30·4 as far as the East coast of England, while the Iceland depression was deeper and was moving East during the day; by the evening of the 30th, the Icelandic depression had moved to the East of Iceland and the area of high pressure had moved a little further to the South-West; by the evening of the 31st, the low had reached the North coasts of Scandinavia, and the high had moved North to the West of the Irish coast; the gradients in the South of England were rather steeper.

On the 28th the wind velocity was small up to 2 kilometres, and thence steadily increased from 2 metres per second to about 23 metres per second at

8 kilometres, the direction between these heights having veered from about 330°
to North. Between 8 and 11·5 kilometres the velocity remained over 20 metres
per second, and reached a maximum of possibly 25 metres per second at 11·7 kilometres; meanwhile the wind had backed from about North to 340°. Above this
point the wind rapidly fell off so that at 13 kilometres it was only 4 metres per
second, and the wind had backed to about 310°. The stratosphere was reached
at 11·5 kilometres with a temperature of −59° C. On the evening of the 29th the
wind was about 6 metres per second up to 2 kilometres, it then increased slowly;
just above 5 kilometres it was 12 metres per second but went down to 8 metres
per second at 6 kilometres, above this there was a slow but steady rise to 24 metres
per second at 11·5 kilometres; above this there was the regular and rapid drop
in wind velocity that had been observed on the previous evening, though the drop
was not quite so rapid, the velocity being 10 metres per second at 13 kilometres.
The wind was a little East of North at almost all heights and was remarkably
constant in direction, but at the greatest heights reached there appeared to be signs
of decided backing. After 57 minutes the balloon was seen to burst and was seen to
fall for some minutes; it must have fallen into the sea and was never recovered. The
stratosphere was reached on this day at 13·7 kilometres at Pyrton Hill, at 13 kilometres at Crinan and Limerick and at 12 kilometres at Glossop.

On July 30th the velocity was about the same as on the previous evening at
ground level, but the increase above was neither so marked nor so regular as on the
preceding days; the maximum was about 16 metres per second at 10·7 kilometres,
and above this it fell to about 11 just below 12 kilometres; the direction, which was
West at ground level, veered more or less regularly to 320° at the greatest height
observed; the sun was not shining on the balloon when it was last seen and
there was some cirrus cloud moving from about 300°. A balloon from Pyrton Hill
was the only one recovered that reached the stratosphere on this day, and the
height of the beginning of the layer was 15·3 kilometres; it is doubtful whether
the balloon sent up at Ditcham was really observed after it had entered the
stratosphere. On the following day, July 31st, the wind being still in the North no
instrument was sent up but a large balloon was watched and was observed till it
attained an estimated height of 13·3 kilometres, when it was seen to burst. The
wind velocity was 5 metres per second at the ground level; at 3 kilometres it was
10 and by 3·5 kilometres it had gone up to 19; between this and 11·5 kilometres
there was a steady increase in velocity (save for a temporary decrease at 7·3 kilometres) to about 35 metres per second at 11·5 kilometres. Above this there was
a marked fall of velocity to 25 at 13·3; the direction showed a steady backing from
about North at ground level to 310° at the highest point. On this day the
stratosphere was reached at Crinan at 11·7 kilometres at 8 a.m., and at 12·5
kilometres at Manchester at 8 p.m.

Taking the July ascents as a whole it thus appears that there is a definite
falling off of wind velocity at a certain height and this height corresponds more
or less with the height of the beginning of the stratosphere.

The next group of ascents in which the balloon was followed into the stratosphere was from Sept. 30th to Oct. 2nd, 1908. On these days the pressure was highest over central Europe and lowest in a depression over Iceland, and in another which during the three days apparently travelled from off the North-West coast of Norway to central Russia. The wind was thus in an almost contrary direction to that of the July ascents. These ascents belong to class (b); in every case the wind increased to many times the gradient velocity, but on Sept. 30th the increase took place slowly, and it was not till a height of 4 kilometres that the observed wind exceeded the gradient.

On Sept. 30th the velocity was small, 3 to 4 metres a second up to 3 kilometres; there was then an increase to 10 metres per second, at which value the velocity remained up to a height of 6·5 kilometres; there was then a steady increase to nearly 30 metres per second at 13·5 kilometres. Above this there is a steady decrease of velocity to 13 metres per second at 16 kilometres. The direction after some irregularities near the surface veered to 210° at 4·5 kilometres; above 10 kilometres it remained constant at about 230° up to 15 kilometres when it backed to 180°. The stratosphere was reached at about 15 kilometres, some distance above the beginning of the decrease of wind velocity, but this balloon was only observed with two theodolites up to 3·5 kilometres, and as it was not seen to burst, the heights towards the end of the ascent must be looked on as only approximate.

On Oct. 1st the balloon was watched till it burst and the highest point deduced from the meteorograph trace was used as the highest point for wind observations; up to 4 kilometres the wind observations were deduced by the two theodolite method; above this the balloon was supposed to have a uniform acceleration up to the highest point reached, and the wind velocities were calculated on the one theodolite method. The velocity was about 3 metres per second at ground level and gradually increased to about 28 metres per second at 12·3 kilometres. The direction was 150° at ground level and slowly veered to 210° at 8 kilometres and then backed to about 185° at 13·5 kilometres. Above that level it varied considerably but the general direction was South; between 14 and 15 kilometres it reached 210°.

On Oct. 2nd the velocity increased from 3 metres per second at ground level to about 16 at 13·5 kilometres; the increase was not very regular, several fairly large fluctuations occurring between ground level and 8 kilometres. Above 13·5 kilometres the velocity fell off to 8 metres per second at 16·2 kilometres. As on the previous day the wind veered with height, being 130° at ground level and 230° at 12 kilometres. Above this there were slow backings and veerings with a rather rapid veer at the highest point.

On these days the same features are shown; there is an increase of velocity up to a certain point followed by a steady decrease at greater heights. The wind direction did not change remarkably, but there are signs, especially on Oct. 1st, of some unsteadiness in direction which is such a remarkable feature of some of the ascents to be considered later.

The stratosphere was remarkably high on these days; the lowest temperatures were found at 16·5 kilometres on Sept. 30th and Oct. 1st and at 14·5 kilometres on Oct. 2nd.

The next ascents in which the stratosphere was reached were on May 6th and 7th (two ascents), 1909. On May 6th after a slight decrease the velocity increased from 10 metres per second at 3 kilometres to 20 metres per second at 11 kilometres, followed by a decrease to 9 metres per second at 12·7 kilometres; the direction meanwhile had veered from 130° to 150°. The stratosphere was reached at about 13 kilometres but there was no sharp point at which it commenced.

On May 7th at 2.52 p.m. the velocity decreased, but there were several fluctuations. Above 11 kilometres there was a steady falling off of velocity to 2 metres per second at 13 kilometres; at 8 kilometres there was a veer of 60°, and there was a further veer of 80° between 12 and 13 kilometres.

The ascent at 6.29 p.m. was very similar to that of the early afternoon; the velocity was rather more irregular and the two veerings took place at a rather lower level; it must however be remembered that the first ascent above 9·5 kilometres and the whole of the later ascent depend on the observations of one theodolite only. The diagrams for the lower of the two veerings are similar (in the two cases) though one was observed by the method of two theodolites, the other by the method of one theodolite. The upper part of the ascent at 6.29 p.m. calls for special attention; the wind direction is seen to vary abruptly from layer to layer, so that no curve can be drawn to show the relation between height and wind direction. That the effect is real and is not due to the errors incident to the one theodolite method is shown by the fact that the bearing of the balloon varies in an irregular manner, increasing and decreasing several times in quick succession; this does not occur when balloons are watched in the lower layers of the atmosphere. The stratosphere was reached at 12·6 kilometres on May 7th. On both days there was a large area of high pressure to the North-East, with a shallow depression to the South-West.

The next high ascent was on Aug. 5th, 1909. The wind velocities were extremely small, averaging about 5 metres per second up to 8 kilometres when a calm layer was met with; above this there was a sharp rise to 10 metres per second at 9 kilometres, accompanied by a change of wind from 85° to 130°; above this the velocity fell off and became extremely irregular, varying from 7 metres per second to almost a calm. It is however in the wind direction that the most striking feature is shown; above 12 kilometres the direction became quite irregular; it is first in one direction and then changes abruptly with height to some quite different direction, many such changes taking place. The trajectory of the balloon makes several loops; the balloon was watched with two theodolites up to 15 kilometres and one of the loops occurred below that height; the loop appears whether the trajectory is computed from the one or the two theodolite method. The rapid and irregular changes in the bearing of the balloon show that towards the end of the ascent, when it was being observed with one theodolite only, the irregularities in direction were real and not due to errors due to the one theodolite method. The distribution of pressure on this day shows a large anticyclone over the whole region, the station being South-West of the central portion.

The ascent of March 3rd, 1910, shows the same features as that of Aug. 5th, 1909. The distribution of pressure was not quite the same; there was a large anticyclone over the Continent and a vast area of low pressure over the Atlantic. The velocities were largest near the ground and up to 2 kilometres, and steadily decreased above; after a preliminary veer there was a steady backing of the wind up to 8 kilometres; at 11 kilometres there was a rapid veer from 20° through South to 270°; above this the direction becomes irregular as in the previous cases; and no curve can be drawn to show the relation between height and wind direction.

It appears from the foregoing cases that the wind usually increases from the ground level and becomes a maximum just below the level of the beginning of the stratosphere; the increase is fairly rapid just below the maximum. Above the maximum the decrease is fairly steady as the balloon enters the stratosphere. This sequence is disclosed in cases of winds from various directions; we have it exemplified in the case of North winds on July 28th, 29th, and 31st, 1908; of West wind on July 30th, 1908; of South winds on Sept. 30th, Oct. 1st and 2nd, 1908; and of an East wind on May 6th, 1909. There are four cases which are different from the above, namely May 7th, 1909 (two ascents), Aug. 5th, 1909, and March 3rd, 1910. In these four cases the wind, which was Easterly at the surface, decreased with height; there is no definite increase in velocity just below the lower limit of the stratosphere. It is in these cases that the wind becomes indeterminate in direction when the stratosphere is entered, as described above. It would seem quite possible that the wind might be shown to behave in a similar manner in all cases if the balloon could be watched to a sufficient height. If there is no real wind in the stratosphere but a strong wind just below friction must cause the air for some distance above the moving current to be carried along in the same direction but with velocity gradually falling off, as is shown in many of the ascents mentioned; the direction should also be in the main the direction of the wind below the stratosphere. Until a height was reached when this "frictional wind" had died out we should not expect to meet with the "stratosphere wind." But when there is little wind in the troposphere we should come to the stratosphere wind directly the balloon entered that layer.

Whether the varying winds, amounting sometimes to 7 or 8 metres per second are real winds blowing in different directions in different levels, or whether they are merely temporary flows it is impossible to say. Light might be thrown on the question by two ascents, one beginning some 10 minutes or so before the other, if both balloons could be watched for a sufficient time.

CHAPTER VIII

RATE OF INCREASE OF WIND VELOCITY NEAR THE SURFACE

IT has been found[1] in a number of cases in the height wind diagrams that the first two points, and sometimes more, lie on a straight line through the origin. That is

$$V_h = \frac{h}{h_0} V_s$$

where V_h is the velocity at the height h above sea level, h_0 the height of a well exposed anemometer above sea level, and V_s the velocity recorded by the anemometer. This linear increase of velocity with height near the ground is called the "regular" increase in the following pages.

At Ditcham there is no anemometer and the horizontal velocity of the balloon in the first minute is taken instead for the purpose of discussing the variation of wind near the surface.

Out of 174 ascents 61 per cent. show the regular increase. The following table shows the percentages of regular to non-regular increase for ascents observed with one and two theodolites.

	No. of cases	Increase regular per cent.	Increase not regular per cent.
One theodolite	124	69	31
Two theodolites	50	42	58
Total	174	61	39

The number of cases of regular increase is much smaller for the two theodolite method than it is for the one theodolite method, and this might seem at first sight to cast some doubt on the reality of the phenomenon. But it must be remembered that the first observation in the two theodolite method is the least satisfactory, both from the large apparent motion of the balloon, owing to its being close to the stations, and from the fact that the triangle made by the balloon with the two ends of the base is not adapted for accurate calculation. It must also be remembered that an error of one or two metres per second would often transfer an ascent from the regular to the not regular class. When this is considered it is a very striking fact that so many of the ascents show this linear increase of velocity with height.

[1] See *Advisory Committee for Aeronautics, Reports and Memoranda*, No. 9, by Dr W. N. Shaw, F.R.S.

If the ascents are classified according to the direction of the surface wind the following percentages are found, taking the North, East, South and West quadrants.

Quadrant	No. of cases	Increase regular per cent.	Increase not regular per cent.
North	47	64	36
East	46	70	30
South	36	44	56
West	33	58	42
Total	162	60	40

In every case there is a preponderance of regular increase except in the South quadrant.

If the North-East, South-East, South-West, and North-West quadrants are taken the following percentages are found.

Quadrant	No. of cases	Increase regular per cent.	Increase not regular per cent.
North-East	50	62	38
South-East	44	64	36
South-West	26	46	54
North-West	41	66	34
Total	161	61	39

If the ascents are classified according to wind velocity the following percentages are found.

Wind velocity metres per second	No. of cases	Increase regular per cent.	Increase not regular per cent.
0 to 4·9	82	67	33
5 to 9·9	84	58	42
Total	166	63	37

It thus appears that the strength of the surface wind does not make much difference to the regularity or otherwise of the initial increase of velocity; but there are fewer cases of regular increase with winds from a Southerly or South-Westerly direction than from other directions. At Ditcham a Southerly wind is one which strikes the line of the South Downs at right angles; the Downs being the first elevated ground met with by such a wind; this would cause irregular eddies over the tops of the hills. The ground also rises 30 to 40 metres to the North of the station from which the balloons are sent up, and the ground to the North is covered with trees some of which are 20 metres high; this must perceptibly affect the air in which the balloon is rising during the first minute of the ascent in the case of Southerly winds.

The formula of increase of velocity with height

$$V_h = \frac{h}{h_0} V_s$$

can only be considered correct for the particular locality under consideration. It is probable that the velocity increases in height by a factor and not by a constant addition, and a formula put forward by Dr Shaw[1] as a likely one is

$$V = \frac{H + a}{a} V_0$$

where H is the height above the ground, V the velocity there, V_0 the anemometer velocity and a a constant which at Ditcham would be 167 metres.

[1] *Advisory Committee for Aeronautics, Reports and Memoranda*, No. 9, p. 8.

CHAPTER IX

General Results; Relation of Vertical Wind Distribution to Surface Pressure Distribution

In considering the various ascents described in the foregoing pages the following facts appear.

The wind distribution of class (a), when the velocity remains fairly constant with regard to height, is found at times when the temperature gradient over the area is not very pronounced. Class (b) on the other hand is found when there is a considerable temperature gradient, the area of low temperatures being more or less coincident with the area of low pressure. This distribution of pressure would, apart from other things, cause an increased pressure gradient in the upper air. Winds of this class are found when cyclonic areas are centred to the North or to the West of the station. This is perhaps the normal condition of affairs when a cyclonic system is passing to the North of these islands, especially in winter, for warm air is drawn from the South-West from the Atlantic, causing a considerable temperature gradient. Winds of class (b) are found (1) to the South of a cyclone that is passing in an Easterly direction to the North of the station, the wind direction being West (Fig. 41), (2) on the North-East of an anticyclone with a depression far to the North-West, the wind direction being Northerly (Fig. 42); (3) on the West side of an anticyclone with a depression far to the North-West, the wind direction being Southerly (Fig. 43). The last type of pressure distribution does not however always give a vertical wind distribution of class (b); it is frequently found associated with class (e 1), the upper wind being North-Westerly. Of the 200 ascents described in this volume 50 are included in class (b), which thus seems to be the commonest form of wind distribution in this country. In fact it seems possible that if observations were carried high enough it would be found that in all cases, save in some cases of Easterly winds, the wind velocity would increase to several times the gradient value before the stratosphere was reached. This leads to the consideration as to whether the classification adopted is a valid one. It must be noted however that in most cases of class (b) the excess of velocity is met with in the first few kilometres. Those cases in which the increase does not take place till a height of 8 to 10 kilometres is reached might perhaps be included in class (a); and it might be better perhaps to differentiate the classes by the behaviour of the wind in the first 5 or 6 kilometres.

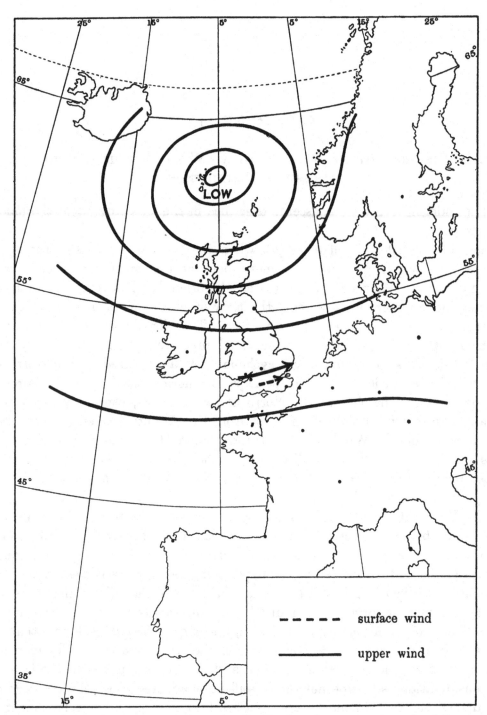

FIG. 41. Relation of Westerly winds to the distribution of pressure at sea level for ascents
in class (b).

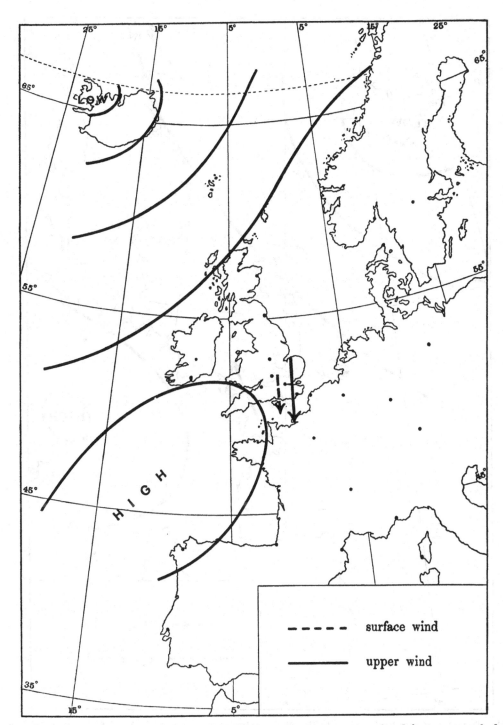

FIG. 42. Relation of Northerly winds to distribution of pressure at sea level for ascents of class (b).

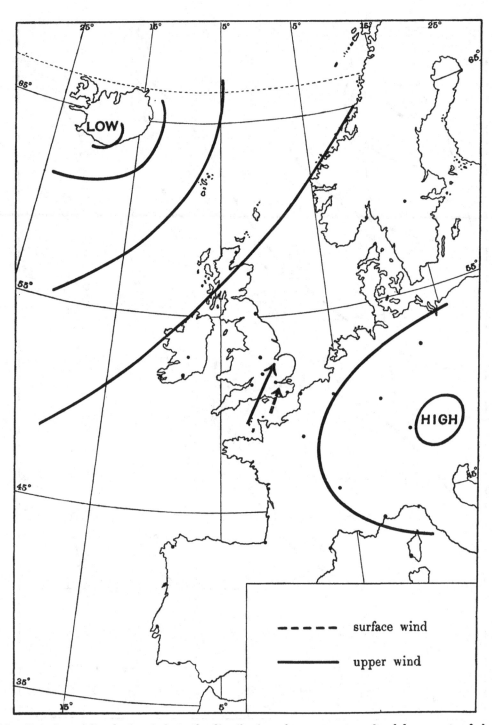

FIG. 43.   Relation of Southerly winds to the distribution of pressure at sea level for ascents of class (b).

Class (c), consisting of winds falling off in velocity with height is restricted almost entirely to Easterly winds, and in general the pressure is high to the North or North-East, and low to the South (Fig. 44). The dividing line between this class and the reversals is not sharply marked; a surface wind from the East often underlies a wind from a Westerly or South-Westerly direction, but this is by no means always the case, and both these classes must be looked at in conjunction with class (e)—the upper winds that blow away from centres of low pressure. Indeed the different classes into which the varying forms of vertical wind distribution have been divided in the foregoing pages must be regarded to a certain extent as artificial; in Nature they must be looked at as being in reality only different parts of the complicated structure of the Atmosphere. When there is no low pressure system for a long way to the West of the station the Easterly wind seems to be maintained, though with diminished velocity, through the whole thickness of the troposphere; but if there is a low pressure system to the West the type is usually of class (d) or (e) with a reversal to a Southerly or to a Westerly wind.

A consideration of the cases in class (e) will I think make it clear that from centres of low pressure we may have vast currents of air flowing in the upper regions to great distances from the centres of the disturbances. The positions in which these conditions occur have already been indicated—on the North-West, North, and North-East of an anticyclone and on the East, South-East, and sometimes on the North-East of a depression. Sometimes on the West of an anticyclone the distribution is of class (b), sometimes of class (e), the difference probably depending on the positions of neighbouring low pressure areas. Further investigation is needed on this and indeed on many other points.

Air rises near the centres of depressions; the air drawn into the cyclonic area and rising near its centre must flow outward in some way, and it appears from these observations that the outflow does not take place symmetrically in every direction, but it forms currents which flow usually in such a direction as to cause Westerly or North-Westerly winds in the upper layers.

It is of the greatest interest to know to what points these upper currents flow. They must ultimately end in some region where they flow downward again towards the surface, for the equivalent of the air which is raised in the depression must return to the surface to take the place of the air that has been raised. Certain parts of anticyclonic areas are presumably those in which the air reaches the surface, and some cases have been indicated, notably those of Nov. 6th to 8th, 1908, where there is distinct evidence of a downward flow of the upper current. It is probably not in the centres of anticyclonic areas that the descent of air takes place. Dr W. N. Shaw, F.R.S., and Mr R. G. K. Lempfert, in their *Life History of Surface Air Currents*, state that they have "failed to identify the central areas of *well marked anticyclones* as regions of origin of air currents," and that the areas of descending air seem to be the shoulders and protuberances of anticyclones and the regions between cyclonic depressions[1].

[1] *The Life History of Surface Air Currents*, p. 24.

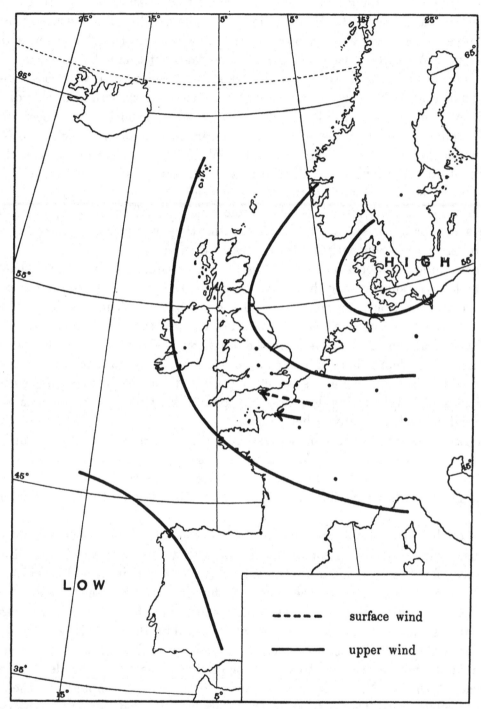

FIG. 44.   Relation of Easterly winds to the distribution of pressure at sea level for ascents in class (c).

Fig. 45 shows diagrammatically a possible way in which the upper winds flow from the centre of a low pressure system towards an anticyclonic area. If the upper wind finds its way to the surface in some part of this latter region it must flow in a clockwise direction round the high pressure centre and return towards the low pressure in the direction indicated on the diagram. Now the air that has risen over the low pressure will usually be cooled at the wet adiabatic rate, while the air that descends over the anticyclone will be warmed at the dry adiabatic rate. If the same air has accomplished the journey it will return towards the anticyclonic area with a higher temperature than that at which it started. This warm air will move towards the East or South-East side of the depression and being a warm current will tend to rise further to the East of the point from which it, or its

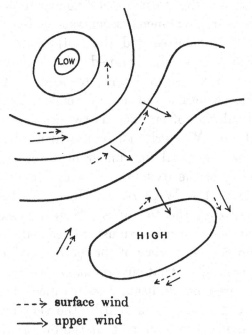

---> surface wind
——> upper wind

FIG. 45. Diagrammatic map showing supposed relations of winds to the isobars.

equivalent, had started. In other words the point at which the air is rising, that is the centre of the depression, will have moved somewhat in the direction from which the anticyclonic air is flowing to feed it. Is it possible that these considerations may throw some light on the progressive motion of a cyclonic system in certain cases?

These considerations may also explain the fact that cyclonic systems may sometimes remain stationary for considerable time. If the air from an anticyclonic region passes over a cold land surface it will be chilled before reaching the low, and will not therefore have a tendency to rise in front of the depression, in other words the depression will not advance as in the case supposed above. This condition may be found in winter or early spring with a high to the North or North-West of these islands and a low over the Bay. The air flowing from the North on the East side of the high is partly drawn from the arctic regions and is also cooled as it flows over the cold regions of the Continent, and the front of the low over the Bay will be partly fed by this cold air; a depression in such conditions may remain stationary for some days (Fig. 46).

In considering rising or falling currents of air it must be remembered that such currents are probably not exactly of the character formerly supposed. It seems likely that the rising air of a cyclonic depression is largely the rising of a warm current of air and its flowing over a colder current from a different direction, and is not merely the ascent of air warmer than its surroundings[1]. The old idea of a warm core above a depression, causing a vertical circulation must be discarded since the temperatures over cyclonic systems have been found to be lower than those over anticyclones. Mr W. H. Dines in a paper read before the Royal Meteorological Society says "the term ascending current of a cyclone...appears to be incorrect. The actual phenomena seem rather to be a bulging upward of the strata between 1 or 2 kilometres and the isothermal, a bulging downward of the strata above the isothermal, accompanied with a lateral contraction over a large area of the lower strata and a lateral expansion of the strata below the isothermal. I much doubt whether the actual vertical motion of any air particle involved in a cyclonic circulation often exceeds two kilometres[2]."

The North-Westerly or Westerly current which the observations of pilot balloons have disclosed as the usual forerunner of an advancing depression should be considered in any theory of the dynamics of a cyclonic system. Is this current composed of the same air as the Westerly or North-Westerly current of the rear of the system which flows over the top of the South-Westerly or Southerly wind of the surface? This and other questions must be decided by future research. This current too is of some importance in view of the methods of forecasting of M. Gabriel Guilbert. M. Guilbert maintains[3] that divergent winds, that is winds which blow away from areas of low pressure, indicate the advance of a low pressure system. A divergent wind in the upper air appears to be the very frequent forerunner of an approaching depression, and it may well be that the upper wind may descend to the surface in places and give a divergent component to the surface wind, as observed by M. Guilbert.

A word must be said of the frequency with which reversals or great changes of wind direction are associated with thunderstorms; again and again as will be seen in the account of the ascents under classes (d) and (e) thunderstorms occur over some part of the country when these conditions are revealed. It is indeed possible that some such conditions are necessary for the production of a thunderstorm. It is difficult to see how the difference of potential between one layer of the atmosphere and another can be maintained unless wind currents from different directions are bringing masses of air at different potentials near together; but if say a polar wind

[1] *Forecasting Weather*, by Dr W. N. Shaw, F.R.S., pp. 211, 212.

[2] "The Statical changes of pressure and temperature in a column of air that accompany changes of pressure at the bottom," *Quarterly Journal Royal Meteorological Society*, Vol. XXXVIII. p. 41.

[3] *Nouvelle Méthode de Prévision du Temps.*

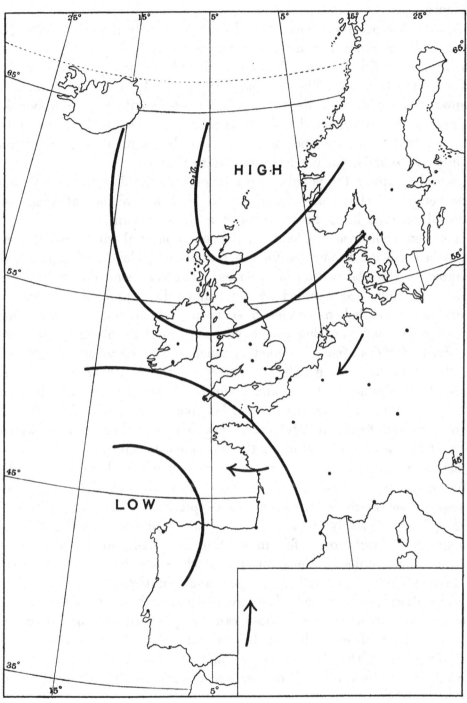

FIG. 46.

is flowing under an equatorial and the air in the two regions is at different potentials a constant difference of potential will be maintained as long as the two opposite winds persist.

A curious fact has been noticed at Ditcham, namely that with light Northerly winds, especially in summer, and when a reversal may be expected in the upper air the guns from ships firing in the Channel are heard far more plainly even than when there is a Southerly wind. The explanation is probably that the sound waves that travel upward in a Northerly direction are refracted on entering the upper Southerly current, and are finally refracted down to the surface. This perhaps accounts for the fact often noticed that guns and fog signals may be heard at a great distance from the origin of the sounds, but not in intermediate places.

It may also explain the popular superstition that the firing of heavy guns brings rain; the conditions that cause the firing to be heard plainly at great distances, being also those that bring heavy rains of a thunderstorm type.

It has been mentioned in Chapter I that an equatorial wind flowing over a polar one may be a condition favourable for the formation of cloud and rain, and therefore unfavourable for the observation of balloons. The very common type of weather of North-Easterly winds accompanied by cloud and rain is probably of this character. On several occasions ballons sondes have been sent up with a North-Easterly or Easterly wind in cloudy weather and have been found at places from North-West to North-East of the station. A good instance of this occurred during the ascents of June 3rd, 1909, when the balloons started to the South-West and were found near Norwich, Maidenhead, Didcot and Oxford. Another example is in the ascents of May 18th to 19th, 1910, when several balloons started in a Westerly direction but were recovered from the Midlands. On May 19th there was extraordinary audibility of the sounds of trains on the Portsmouth railway.

It has been noticed at Ditcham on occasions when there is a Southerly or South-Westerly current flowing over a North-Easterly one that the trace of the microbarograph shows marked fluctuations of pressure. So much is this the case that with a North-Easterly wind with low clouds and rain and fluctuations of the microbarograph no hesitation is felt in sending up ballons sondes although the sea coast lies not many miles to the South; in such cases the balloons almost always enter a reverse current and fall in the midland counties.

Further observations of this class are needed, and observations of the upper clouds would be instructive when these can be glimpsed through temporary rifts in the lower cloud sheet. The conditions will often be found to be those of class (e.2) (Fig. 47) with a low pressure system to the South-West, and an upper wind blowing from the centre of the depression; this seems to have been the case on June 3rd, 1909.

Further observations of pilot balloons are greatly to be desired, and if observations could be made from several stations simultaneously much more might be learned about the movements of the atmosphere. Four stations in the British Islands, say one in the South of England, one in the North-East of England or

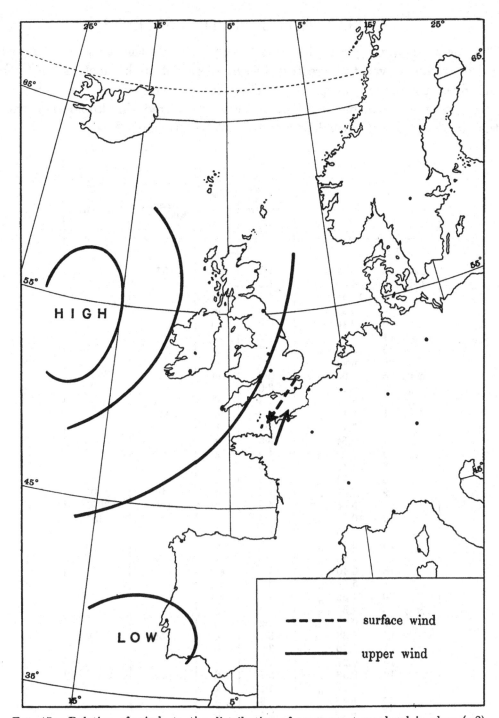

FIG. 47.   Relation of winds to the distribution of pressure at sea level in class (e 2).

the South-East of Scotland, one in the South-West of Ireland, and one in the West or the North-West of Scotland, at which observations of pilot balloons should be made on every day when conditions allowed would probably result in a very great addition to our knowledge of the structure of the atmosphere.

GENERAL TABLE OF THE ASCENTS.

The first column gives the date, the second the time of the ascent. The third column gives the greatest height observed in kilometres. The fourth column gives the class (see p. 35). The fifth column gives the point of fall in those cases in which the balloon was recovered in kilometres and the direction in degrees from North through East, South and West.

Day and hour of ascent			Greatest observed height. Kilometres	Class	Point of fall	
					Distance. Kilometres	Direction in degrees from N.
1907						
Jan.	25	11.42 a.m.	1·5	d	37	90°
Feb.	1	4.6 p.m.	1·9	a		
,,	2	11.45 a.m.	1·2	m		
,,	4	10.44 ,,	1·2	m		
,,	8	0.45 p.m.	0·9	e l		
,,	14	2.9 ,,	1·2	m		
,,	23	3.43 ,,	1·3	m		
,,	26	noon	1·2	c		
Mar.	28	5.55 p.m.	2·0	m	137	360°
,,	29	6.25 ,,	3·0	a	309	10°
,,	30	6.19 ,,	1·3	d		
Apr.	1	6.12 ,,	3·9	d	39	1°
,,	2	5.46 ,,	2·2	b		
,,	4	6.34 ,,	3·5	e		
,,	5	4.40 ,,	4·0	a		
,,	6	6.13 ,,	4·8	a		
,,	8	6.14 ,,	3·3	a		
,,	9	6.20 ,,	1·6	d	14·5	160°
,,	13	5.56 ,,	5·9	b	169	300°
,,	15	10.50 a.m.	1·6	d	7	315°
,,	15	6.36 p.m.	1·5	d		
,,	16	6.27 ,,	1·7	c		
,,	17	4.49 ,,	1	a		
,,	18	6.40 ,,	5	e l		
,,	19	5.59 ,,	3	e l		
,,	20	4.52 ,,	4	e l		
May	3	2.33 ,,	1·2	b		
[1] ,,	10	10.22 a.m.	2·2	b		
[1] ,,	10	6.32 p.m.	3·3	b	(150	22°)
[1] ,,	11	5.40 ,,	3·2	b		
[1] ,,	13	6.27 ,,	1·0	c		
[1] ,,	14	6.53 ,,	2·2	d	(40	260°)
[1] ,,	16	7.17 ,,	4·2	a		
[1] ,,	17	6.43 ,,	1·9	d		
[1] ,,	18	6.53 ,,	2·4	b		
[1] ,,	20	4.49 ,,	1·5	m		
[1] ,,	21	6.8 ,,	3	d		
[1] ,,	24	7.3 ,,	7·0	b		
[1] ,,	25	7.13 ,,	2·7	d		
[1] ,,	27	6.11 ,,	6·5	d		
[1] ,,	29	7.11 ,,	1·5	c		

[1] Totland Bay.

Day and hour of ascent	Greatest observed height. Kilometres	Class	Point of fall	
			Distance. Kilometres	Direction in degrees from N.
1907				
June 6 3.9 p.m.	2·2	b		
,, 6 6.38 ,,	3·7	b		
,, 8 7.15 ,,	5·5	b		
,, 17 7.15 ,,	1·6	m		
,, 20 6.40 ,,	2·4	b	106	55°
,, 29 7.11 ,,	2·5	d		
*¹July 1 3.15 ,,	1·0	d	(66	75°)
¹ ,, 1 4.7 ,,	2·1	d		
,, 22 0.47 ,,	1·8	d		
,, 24 3.7 ,,	3·5	e 2		
Aug. 5 2.24 ,,	1·2	d	35	31°
,, 5 7.19 ,,	2·7	b		
,, 23 7.5 ,,	3·5	d		
,, 28 11.38 a.m.	2·5	d		
,, 28 6.23 p.m.	6·0	d		
Sept. 1 10.22 a.m.	4·8	b		
* ,, 4 5.4 p.m.	2·0	e 1	6·5	73°
,, 23 2.17 ,,	7·8	e 1	9	75°
,, 27 4.38 ,,	4·3	b	46	309°
Oct. 14 5.24 ,,	1·0	d		
Dec. 26 0.4 ,,	4·2	c	24·5	270°
1908				
Jan. 2 10.43 a.m.	4·2	c		
,, 2 3.47 p.m.	2·2	c		
,, 3 10.49 a.m.	1·5	c	64	268°
,, 3 11.19 ,,	2·7	c	61	267°
,, 3 3.59 p.m.	5·3	c		
,, 4 11.2 a.m.	1·4	c		
,, 11 3.38 p.m.	6·3	e 1	10·5	240°
,, 17 2.25 ,,	2·8	a		
Feb. 1 4.32 ,,	2·0	a		
,, 2 4.6 ,,	3·7	b		
,, 4 4.48 ,,	2·5	b		
,, 5 10.20 a.m.	2·0	b		
,, 8 11.31 ,,	2·9	a		
,, 8 4.25 p.m.	3·4	a		
,, 12 2.35 ,,	6·1	b	55	358°
,, 13 3.8 ,,	3·8	b		
,, 14 4.38 ,,	5·2	e 1		
,, 26 10.28 a.m.	2·3	b	18·5	130°
,, 26 11.5 ,,	2·2	b	14·5	112°
Mar. 3 10.12 ,,	3·6	d	13	53°
,, 5 11.30 ,,	3·4	e 1		
,, 5 5.5 p.m.	2·0	e 1	11	60°
,, 11 5.28 ,,	2·6	b		
,, 12 5.52 ,,	3·2	d	8·5	164°
,, 13 5.40 ,,	2·0	d		
,, 14 10.37 a.m.	1·7	c	3	315°?
,, 21 5.50 p.m.	5·0	d		
,, 24 10.19 a.m.	4·7	e 1	8·5	112°
,, 27 11.4 ,,	2·5	e 1		
,, 27 5.41 p.m.	3·3	c		
,, 28 5.40 ,,	3·0	e 1		

* Ballon sonde.   ¹ Chobham Common.

Day and hour of ascent			Greatest observed height. Kilometres	Class	Point of fall	
					Distance. Kilometres	Direction in degrees from N.
1908						
Mar.	30	11.29 a.m.	1·1	m	9	65°
Apr.	1	5.50 p.m.	2·0	e 1		
,,	1	6.13 ,,	1·3	e 1		
,,	4	4.14 ,,	2·0	m		
,,	8	10.39 a.m.	4·2	a		
,,	8	6.46 p.m.	4·7	a		
,,	9	11.8 a.m.	3·9	c	11·5	180°
,,	9	5.53 p.m.	4·8	d	294	120°
,,	16	11.12 a.m.	2·5	c	89	268°
,,	29	3.57 p.m.	6·2	e 1		
May	2	0.48 ,,	6·7	e 2	21	35°
,,	11	0.37 ,,	3·9	b		
,,	18	6.37 ,,	7·0	a	114	80°
,,	21	7.18 ,,	3·0	b		
,,	23	11.0 a.m.	3·0	a		
,,	27	10.16 ,,	3 6	b	30	210°
,,	29	8.12 ,,	2·5	c		
,,	30	7.2 p.m.	7·0	a	56	300°
June	1	7.32 ,,	4·6	b	97	333°
,,	2	7.15 ,,	6·3	d	66	21°
,,	3	10.19 a.m.	4·9	e 2		
,,	3	6.54 p.m.	11·6	e 2		
,,	4	7.0 ,,	1·8	c		
,,	5	6.54 ,,	5·2	b		
,,	10	7.15 ,,	6·4	e 1		
,,	22	6.36 ,,	13·0	e 1	45	212°
,,	23	7.0 ,,	3·7	a		
,,	26	7.18 ,,	5·9	a		
*July	27	7.19 ,,	8·6	d	183	44°
,,	28	7.0 ,,	13·0	e 1 ; f	145	170°
,,	29	9.27 a.m.	4·4	a		
,,	29	5.3 p.m.	5·3	a		
* ,,	29	7.0 ,,	13·3	e 1 ; f		
,,	30	8.6 a.m.	5·4	e 1		
,,	30	0.36 p.m.	4·0	e 1		
* ,,	30	7.0 ,,	11·8	e 1 ; f		
,,	31	7.0 ,,	13·4	e 1 ; f		
Aug.	1	8.10 a.m.	3·4	e 1		
,,	1	5.53 p.m.	4·2	e 1		
,,	1	7.5 ,,	4·0	e 1		
,,	2	5.54 ,,	4·1	e 1		
,,	2	7.30 ,,	4·8	e 1		
Sept.	30	8.36 a.m.	3·7	d		
,,	30	11.17 ,,	6·7	d		
* ,,	30	4.31 p.m.	16·0	b ; f	105	40°
Oct.	1	8.15 a.m.	3·6	b		
* ,,	1	4.20 p.m.	17·6	b ; f	114	18°
,,	2	8.20 a.m.	3·9	b	45	11°
* ,,	2	4.20 p.m.	16·2	b ; f	57	10°
,,	3	10.45 a.m.	3·2	b		
,,	3	3.25 p.m.	4·2	b		
Nov.	3	0.47 ,,	6·5	d		
,,	6	10.59 a.m.	9·5	d		
,,	7	3.25 p.m.	6·1	d		

* Ballon sonde.

Day and hour of ascent		Greatest observed height. Kilometres	Class	Point of fall	
				Distance. Kilometres	Direction in degrees from N.
1908					
Nov. 8	4.26 p.m.	4·5	d	40	266°
,, 16	10.47 a.m.	5·5	e 1		
1909					
Jan. 12	10.58 a.m.	0·9	b		
,, 12	2.22 p.m.	2·1	b		
* ,, 12	3.54 ,,	3·2	b		
,, 15	11.40 a.m.	3·0	b		
,, 19	10.34 ,,	2·0	m		
,, 19	0.33 p.m.	2·9	m		
,, 20	10.20 a.m.	2·9	c		
,, 30	3.21 p.m.	3·0	b		
Feb. 5	4.28 ,,	3·5	b		
,, 6	4.48 ,,	1·7	b		
,, 7	4.28 ,,	4·1	e 1	11·5	326
,, 12	5.8 ,,	4·2	c		
,, 13	5.6 ,,	3·6	b		
,, 14	4.50 ,,	3·5	b		
,, 15	2.21 ,,	5·0	b		
,, 17	8.17 a.m.	6·0	e 1		
,, 18	4.43 p.m.	7·9	e 1		
,, 19	10.2 a.m.	2·6	c	39	309°
,, 19	4.44 p.m.	6·4	a		
,, 20	10.26 a.m.	5·0	a	56	352°
,, 20	4.46 p.m.	6·0	a	37	353°
,, 21	2.35 ,,	5·6	c		
,, 22	11.38 a.m.	3·3	c		
,, 22	4.52 p.m.	5·8	a	35	231°
,, 23	4.30 ,,	3·7	a		
,, 24	4.45 ,,	5·5	a	36	225°
Mar. 5	5.13 ,,	4·5	e 1		
Apr. 19	8.15 a.m.	4·7	b	39	13°
,, 21	7.58 ,,	5·2	a		
,, 26	2.41 p.m.	2·9	b	87	29°
May 2	7.7 ,,	5·0	e 1		
,, 4	8.15 a.m.	3·0	b		
,, 4	7.4 p.m.	5·5	c		
,, 5	8.10 a.m.	2·4	c	20	282°
,, 5	2.54 p.m.	6·3	a		
* ,, 5	6.43 ,,	10·2	a	94	292°
,, 6	10.33 a.m.	2·8	c	197	310°
,, 6	0.27 p.m.	8·0	a		
* ,, 6	6.25 ,,	12·8	a ; f	81	305°
,, 7	2.52 ,,	13·0	c ; f	41	290°
* ,, 7	6.29 ,,	15·0	c ; f	46	272°
,, 16	6.22 ,,	3·0	e 2		
,, 31	0.13 ,,	4·3	b		
June 1	10.23 a.m.	2·4	d		
,, 3	0.5 p.m.	1·2	c		
,, 3	0.59 ,,	3·6	e 2		
,, 21	7.0 ,,	4·2	b		
Aug. 5	0.8 ,,	2·8	a		
,, 5	2.30 ,,	8·0	a		
* ,, 5	6.33 ,,	18·5	a ; f	19	254°
1910					
*Mar. 3	4.30 p.m.	15·2	c ; f	20	265°

* Ballon sonde.

TABLE OF THE ASCENTS GIVING THE WIND VELOCITIES IN METRES PER SECOND AND THE WIND DIRECTIONS IN DEGREES FROM NORTH (THROUGH EAST, SOUTH, AND WEST) FOR EVERY HALF KILOMETRE.

Figures in square brackets indicate that the height opposite the figures has been nearly but not quite reached.

All ascents are from Ditcham and observed with one theodolite unless otherwise stated.

The cause of cessation of observations when it has been recorded, is given by the words distance, clouds, etc., in the notes at the foot of the column of observations.

CLASS (a).

"Solid" current; little change in velocity or direction, wind reaches the gradient value and does not increase very much at greater altitudes.

	Feb. 1, 1907 4.6 p.m.		Mar. 29, 1907 6.25 p.m.		April 5, 1907 4.40 p.m.		April 6, 1907 6.13 p.m.		April 8, 1907 6.14 p.m.		April 17, 1907 4.49 p.m.		
	vel.	dir.	vel.	dir.	vel.	dir.	vel.	dir.	vel.	dir.	vel.	dir.	
Gradient	—	—	6·0	180	11·2	220	16·0	270	11·2	280	8·0	40	Gradient
Surface	7·3	10	6·0	130	4·5	235	6·0	275	6·5	250	5·0	355	Surface
0·5	8·5	25	8·5	175	9·5	245	11·5	270	12·0	260	9·5	330	0·5
1·0	9·8	25	8·5	185	9·5	230	16·0	265	14·5	280	8·5	345	1·0
1·5	9·0	40	9·7	190	6·2	225	19·5	260	12·5	290	8·0	350	1·5
2·0	9·0	10	9·2	205	7·5	220	15·0	260	9·0	290	9·0	350	2·0
2·5			11·0	185	7·0	225	14·5	255	9·0	280			2·5
3·0			11·0	190	11·0	230	14·5	260	11·5	275			3·0
3·5					12·0	225	14·5	255					3·5
4·0					9·5	265	14·5	260					4·0
	distance												

CLASS (a)—continued.

Gradient	May 16, 1907 7.17 p.m.		Jan. 17, 1908 2.25 p.m.		Feb. 1, 1908 4.23 p.m.		Feb. 8, 1908 11.31 a.m.		Feb. 8, 1908 4.25 p.m.		April 8, 1908 10.39 a.m.		Gradient
	vel.	dir.	vel.	dir.	vel.	dir.	vel.	dir.	vel.	dir.	vel.	dir.	
Gradient	9·0	350	22·0	240	12·0	355	9·5	315	8·0	320	—	—	Gradient
Surface	4·5	360	—	—	6·5	350	5·5	315	4·5	305	6·5	10	Surface
0·5	9·0	355	25·0	235	10·5	350	7·0	315	6·5	330	6·5	25	0·5
1·0	9·0	335	21·0	230	14·0	350	6·0	325	8·5	335	7·0	35	1·0
1·5	9·5	340	14·5	230	16·0	355	8·5	325	7·0	320	5·0	15	1·5
2·0	8·0	320	23·0	220	16·5	360	9·0	325	11·0	305	7·0	25	2·0
2·5	8·0	320	23·0	220			9·0	320[1]	11·5	330	10·0	40	2·5
3·0	10·0	310					[10·0	315]	10·5	320	7·5	40	3·0
3·5	9·5	315									7·0	60	3·5
4·0	13·0	310									9·5	55	4·0
	Totland Bay						2 theodolites to 2·4 km.		cloud				

[1] Decrease of velocity to 6·0 between 2 km. and 2·5 km.

Gradient	April 8, 1908 6.46 p.m.		May 18, 1908 6.37 p.m.		May 23, 1908 11.0 a.m.		May 30, 1908 7.2 p.m.		June 23, 1908 7.0 p.m.		June 26, 1908 7.18 p.m.		Gradient
	vel.	dir.	vel.	dir.	vel.	dir.	vel.	dir.	vel.	dir.	vel.	dir.	
Gradient	—	—	12·0	270	—	—	6·0	90	6·0	330	10·0	90	Gradient
Surface	5·0	30	3·0	270	5·0	315	3·0	80	4·5	70	4·0	85	Surface
0·5	7·5	25	6·0	300	7·0	320	4·0	90	4·0	30	9·0	60	0·5
1·0	8·0	20	9·5	325	5·5	355	5·0	130	6·0	355	9·5	35	1·0
1·5	9·5	25	5·0	310	5·5	355	5·0	95	7·5	360	10·0	55	1·5
2·0	9·5	20	8·5	270	6·0	5	4·5	110	7·5	15	12·0	60	2·0
2·5	8·5	360	10·0	270	7·0	355	6·0	105	5·5	5	7·5	60	2·5
3·0	7·5	25	9·5	260	[8·5	15]	7·5	105	5·0	20	7·5	70	3·0
3·5	6·0	360	9·5	255			7·5	135	6·0	15	9·0	75	3·5
4·0	3·5	350	9·0	255			5·0	130			13·0	80	4·0
4·5	6·0	350	10·0	270			4·0	145			12·5	85	4·5
5·0			10·5	255			4·0	140			12·0	80	5·0
5·5			13·0	240			6·0	120			11·0	85	5·5
6·0			11·0	245			8·0	135			[8·5	90]	6·0
6·5			9·5	240			7·0	135					6·5
7·0			12·5	255			[7·5	125]					7·0
	2 theodolites to 3 km.				2 theodolites		cloud						

Class (a)—continued.

Gradient	July 29, 1908 9.27 a.m.		July 29, 1908 5.3 p.m.		Feb. 19, 1909 4.44 p.m.		Feb. 20, 1909 10.26 a.m.		Feb. 20, 1909 4.46 p.m.		Feb. 22, 1909 4.52 p.m.		Gradient
	vel.	dir.	vel.	dir.	vel.	dir.	vel.	dir.	vel.	dir.	vel.	dir.	
Gradient	—	30?	4·0	360	12·0	155	10·0	160	12·0	170	—	—	Gradient
Surface	5·5	30	3·0	355	6·5	105	8·0	135	5·0	115	5·0	80	Surface
0·5	8·0	15	3·5	5	12·0	130	9·5	140	7·0	150	5·0	70	0·5
1·0	6·5	330	4·0	25	17·5	145	14·0	160	8·5	150	3·5	85	1·0
1·5	6·0	20	3·0	355	17·0	140	10·0	160	9·0	155	6·5	100	1·5
2·0	5·5	15	7·0	5	13·0	150	9·0	175	7·0	165	5·0	80	2·0
2·5	6·0	20	7·0	10	11·0	145	9·0	160	7·0	160	7·0	65	2·5
3·0	9·0	20	6·5	10	10·5	145	9·5	175	8·5	165	7·0	35	3·0
3·5	8·0	5	6·0	20	10·0	145	11·0	175	9·5	160	8·0	40	3·5
4·0	5·0	15	7·0	360	10·0	145	11·0	155	10·5	185	8·0	40	4·0
4·5			6·5	355	10·0	145	10·5	170	10·0	195	9·5	35	4·5
5·0			8·5	360	8·0	145	11·0	170	8·0	185	9·0	30	5·0
5·5					7·5	140			9·0	190	8·5	15	5·5
6·0					8·5	130			9·0	185	[7·0	25]	6·0
6·5					[10·0	135]							6·5
					2 theodolites to 5·1 km.		2 theodolites to 4·7 km.		distance		2 theodolites burst?		

Class (a). Subclass no current.

Gradient	Aug. 5, 1909 0.8 p.m.		Aug. 5, 1909 2.30 p.m.	
	vel.	dir.	vel.	dir.
Gradient	—	—	—	—
Surface	1·0	60	2·0	210
0·5	1·5	110	3·0	175
1·0	2·0	50	5·0	40
1·5	2·0	60	2·0	70
2·0	3·5	50	4·0	75
2·5	3·5	30	4·5	40
3·0			4·0	30
3·5			3·0	70
4·0			3·0	55
4·5			2·0	40
5·0			3·0	50
5·5			2·0	80
6·0			3·0	60
6·5			2·5	55
7·0			2·5	20
7·5			1·0	100
8·0			[1·0	160]

CLASS (a)—continued.

Gradient	Feb. 23, 1909 4.30 p.m. vel. 11·0?	dir. 70?	Feb. 24, 1909 4.45 p.m. vel. 9·0	dir. 90	April 21, 1909 7.58 a.m. vel. 9·0	dir. 170	May 5, 1909 2.54 p.m. vel. 17·0	dir. 100	May 5, 1909 6.43 p.m. vel. 19·0	dir. 120	May 6, 1909 0.27 p.m. vel. 21·0	dir. 120	Gradient
Surface	6·5	15	5·0	70	6·5	110	14·0	65	—	—	5·0	85	Surface
0·5	10·0	35	8·0	55	8·5	130	13·0	65	—	—	6·0	90	0·5
1·0	9·0	40	7·5	35	3·5	160	20·0	90	15·0	95	13·0	110	1·0
1·5	11·0	35	7·0	55	6·0	140	15·0	110	16·0	105	12·5	115	1·5
2·0	11·5	30	5·5	25	7·0	150	14·5	120	17·5	115	11·0	125	2·0
2·5	11·5	25	3·5	40	7·5	130	16·0	110	17·5	105	12·5	130	2·5
3·0	12·0	25	6·5	35	8·0	135	16·0	110	17·0	110	12·0	130	3·0
3·5	11·5	20	7·0	40	8·5	115	16·0	125	12·0	105	8·5	125	3·5
4·0			9·0	55	7·0	125	17·0	125	15·0	110	10·0	125	4·0
4·5			9·0	45	9·0	145	16·5	130	18·5	120	11·0	130	4·5
5·0			10·5	50	10·0	160	17·0	130	18·0	125	13·0	125	5·0
5·5			10·5	45			17·0	130	17·0	130	12·0	120	5·5
6·0							17·0	130	14·0	125	10·0	130	6·0
6·5									13·0	115	11·0	125	6·5
7·0									14·0	105	12·0	130	7·0
7·5									14·0	125	14·0	125	7·5
8·0									15·0	130	18·5	125	8·0
8·5									19·0	125			8·5
9·0									15·0	130			9·0
9·5									13·0	130			9·5
10·0									17·0	110			10·0
			burst		burst		2 theodolites to 2·4 km.						

CLASS (b). Considerable increase in velocity.

Gradient	April 2, 1907 5.46 p.m.		April 13, 1907 5.56 p.m.		May 3, 1907 2.33 p.m.		May 10, 1907 10.22 a.m.		May 10, 1907 6.32 p.m.		May 11, 1907 5.40 p.m.		Gradient
	vel.	dir.	vel.	dir.	vel.	dir.	vel.	dir.	vel.	dir.	vel.	dir.	
Gradient	11·5	180	7·0	130	20·0	255	12·5	205	9·0	200	4·0	190	Gradient
Surface	3·5	130	6·0	85	5·5	260	3·0	180	2·5	105	2·0	90	Surface
0·5	10·0	185	6·0	90	12·0	260	9·0	205	12·5	175	5·0	175	0·5
1·0	15·0	185	6·0	135	20·0	265	10·0	205	14·0	185	12·0	165	1·0
1·5	19·0	195	7·0	165			17·0	200	17·0	190	11·0	185	1·5
2·0	22·0	195	10·0	160			19·0	200	15·5	190	15·0	190	2·0
2·5			10·0	130					15·0	190	19·5	190	2·5
3·0			9·0	125					20·0	185	19·0	190	3·0
3·5			7·0	140									3·5
4·0			12·0	140									4·0
4·5			15·0	140									4·5
5·0			18·0	145									5·0
5·5			16·0	140									5·5
6·0			[17·0	140]			Totland Bay		cloud Totland Bay		Totland Bay		6·0

Gradient	May 18, 1907 6.53 p.m.		May 24, 1907 7.3 p.m.		June 6, 1907 3.9 p.m.		June 6, 1907 6.38 p.m.		June 8, 1907 7.15 p.m.		June 20, 1907 6.40 p.m.		Gradient
	vel.	dir.	vel.	dir.	vel.	dir.	vel.	dir.	vel.	dir.	vel.	dir.	
Gradient	6·0	30	7·0	150	14·0	280	15·0	270	9·0	210	12·0	250	Gradient
Surface	4·5	10	3·5	235	9·0	285	6·5	280	2·5	180	4·5	225	Surface
0·5	10·5	20	9·0	230	13·0	280	13·0	285	6·0	180	18·0	240	0·5
1·0	12·0	25	12·0	230	20·0	280	13·5	295	12·0	205	20·0[1]	240	1·0
1·5	10·0	20	10·0	230	20·0	270	13·5	295	16·0	215	19·0	240	1·5
2·0	8·0	25	10·0	205	20·0	275	14·0	300	11·5	215	24·0	235	2·0
2·5			10·0	205			16·0	305	12·0	225	[30·0	235]	2·5
3·0			11·0	180			19·0	305	11·0	215			3·0
3·5			9·5	175			21·5	310	14·0	220			3·5
4·0			10·5	175					17·0	230			4·0
4·5			12·5	180					16·0	240			4·5
5·0			13·5	175					18·0?	235			5·0

[1] vel. about 25·0 at 1·7 km.

Continued on next page.

CLASS (b)—continued.

	May 18, 1907 6.53 p.m.	May 24, 1907 7.3 p.m.		June 6, 1907 3.9 p.m.	June 6, 1907 6.38 p.m.	June 8, 1907 7.15 p.m.		June 20, 1907 6.40 p.m.	
		vel.	dir.			vel.	dir.		
5·5		15·0	175			17·0	240		5·5
6·0		14·0	180						6·0
6·5		16·0	180						6·5
7·0		18·0	190						7·0
	Totland Bay cloud	Totland Bay distance		distance		lost behind house			

	Aug. 5, 1907 7.19 p.m.		Sept. 1, 1907 10.22 a.m.		Sept. 27, 1907 4.38 p.m.		Feb. 2, 1908 4.6 p.m.		Feb. 4, 1908 4.48 p.m.		Feb. 5, 1908 10.20 a.m.		
	vel.	dir.	vel.	dir.	vel.	dir.	vel.	dir.	vel.	dir.	vel.	dir.	
Gradient	8·0	310	4·0	360	9·0	100	7·5	355	11·0	5	11·5	360	Gradient
Surface	2·0	245	5·5	25	6·5	90	5·5	350	7·0	355	2·0	360	Surface
0·5	3·0	290	4·5	20	11·5	115	8·5	355	13·0	355	5·0	15	0·5
1·0	2·5	315	4·5	290	15·0	130	9·0	30	13·0	10	10·0	25	1·0
1·5	3·0	240	5·0	300	14·0	140	9·5	25	15·0	20	10·0	25	1·5
2·0	12·0	215	7·0	295	13·0	140	17·0	15	19·5	20	13·0	25	2·0
2·5	13·0	205	8·0	290	12·0	145	22·5	25					2·5
3·0			13·0	285	13·0	140	27·5	30					3·0
3·5			16·0	280	15·0	125	28·0	25					3·5
4·0			16·0	270	16·0	130							4·0
4·5			19·5	265									4·5
	cloud				2 theodolites to 1·8 km.						lost owing to quick movement		

	Feb. 12, 1908 2.35 p.m.		Feb. 13, 1908 3.8 p.m.		Feb. 26, 1908 10.28 a.m.		Feb. 26, 1908 11.5 a.m.		Mar. 11, 1908 5.28 p.m.		May 11, 1908 0.37 p.m.		
	vel.	dir.	vel.	dir.	vel.	dir.	vel.	dir.	vel.	dir.	vel.	dir.	
Gradient	8·0	150	7·0	190	12·5	295	12·5	295	5·0	320	—	—	Gradient
Surface	4·0	135	4·0	170	7·5	295	6·5	295	5·0	310	3·0	145	Surface
0·5	7·0	135	5·0	250	9·5	310	7·5	300	6·0	325	2·5	145	0·5
1·0	6·5	170	9·0	270	6·0	295	8·5	290	5·5	345	6·0	175	1·0

Continued on next page.

C.   12

CLASS (b)—continued.

	Feb. 12, 1908 2.35 p.m.		Feb. 13, 1908 3.8 p.m.		Feb. 26, 1908 10.28 a.m.		Feb. 26, 1908 11.5 a.m.		Mar. 11, 1908 5.28 p.m.		May 11, 1908 0.37 p.m.		
	vel.	dir.	vel.	dir.	vel.	dir.	vel.	dir.	vel.	dir.	vel.	dir.	
1·5	5·0	145	6·0	230	9·0	295	9·0	285	8·0	350	6·0	235	1·5
2·0	6·0	200	7·5	210	15·0	330	15·0	320	10·5	340	7·0	220	2·0
2·5	9·0	185	11·0	185	[18·5	325]			11·0	315	9·5	215	2·5
3·0	9·0	180	15·0	210							10·0	205	3·0
3·5	8·5	170	17·0	190							10·0	195	3·5
4·0	8·0	185									[19·0?	195]	4·0
4·5	11·5	190											4·5
5·0	13·0	190											5·0
5·5	14·5	185											5·5
6·0	13·0	180											6·0
	2 theodolites to 2·2 km. distance		2 theodolites to 2·2 km.		2 theodolites burst		2 theodolites burst		2 theodolites cloud		cloud		

	May 21, 1908 7.18 p.m.		May 27, 1908 10.16 a.m.		June 1, 1908 7.32 p.m.		June 5, 1908 6.54 p.m.		Oct. 1, 1908 8.15 a.m.		Oct. 2, 1908 8.20 a.m.		
	vel.	dir.	vel.	dir.	vel.	dir.	vel.	dir.	vel	dir.	vel.	dir.	
Gradient	12·0	300	5·0	65	2·0	150	6·0	360	6·0	180	4·0	180	Gradient
Surface	7·5	290	2·0	55	3·5	70	4·5	325	3·0	30	2·5	100	Surface
0·5	10·5	295	3·5	50	3·5	120	8·5	335	5·0	75	5·0	140	0·5
1·0	11·5	265	5·5	45	6·5	160	6·0	330	4·0	90	5·5	170	1·0
1·5	10·0	250	8·5	35	12·0	170	7·0	305	4·5	75	8·0	170	1·5
2·0	10·0	240	8·5	30	11·5	160	11·5	315	7·0	80	3·0	165	2·0
2·5	14·0	235	8·5	30	12·0	150	13·0	290	9·5	90	5·0	170	2·5
3·0	25·0	230	5·0	15	17·0	145	12·0	300	8·5	85	9·0	180	3·0
3·5			8·5	15	18·0	150	13·5	300	12·0	90	10·0	170	3·5
4·0					18·0	155	16·0	300			8·5	185	4·0
4·5					15·0	155	15·0	295					4·5
5·0							12·5	290					5·0
							distance						

CLASS (b)—continued.

	Oct. 3, 1908 10.45 a.m.		Oct. 3, 1908 3.25 p.m.		Jan. 12, 1909 10.58 a.m.		Jan. 12, 1909 2.22 p.m.		Jan. 12, 1909 3.50 p.m.		Jan. 15, 1909 11.40 a.m.		
	vel.	dir.	vel.	dir.	vel.	dir.	vel.	dir.	vel.	dir.	vel.	dir.	
Gradient	7·0	165	5·0	180	—	—	12·0	285	12·5	290	14·0	270	Gradient
Surface	2·0	115	2·5	125	8·0	270	8·0	290	6·0	295	4·5	270	Surface
0·5	7·0	130	5·5	160	16·0	280	10·0	295	10·0	290	7·5	275	0·5
1·0	12·5	155	9·0	175	[20·5	290]	15·0	305	12·0	300	14·0	285	1·0
1·5	9·5	150	10·0	165			20·0	295	17·5	300	14·0	275	1·5
2·0	11·5	155	7·5	175			29·0	300	24·0	300	14·0	270	2·0
2·5	8·5	160	8·0	175					26·0	300	17·0	280	2·5
3·0	7·0	155	7·0	175					24·0	295	26·0	270	3·0
3·5			8·5	170									3·5
4·0			12·0	165									4·0
							distance		2 theodolites to 2·5 km.				

	Jan. 30, 1909 3.21 p.m.		Feb. 5, 1909 4.28 p.m.		Feb. 6, 1909 4.48 p.m.		Feb. 13, 1909 5.6 p.m.		Feb. 14, 1909 4.50 p.m.		Feb. 15, 1909 2.21 p.m.		
	vel.	dir.	vel.	dir.	vel.	dir.	vel.	dir.	vel.	dir.	vel.	dir.	
Gradient	11·0	350	18·0	330	5·0	360	11·0	65	—	—	10·0	360	Gradient
Surface	7·5	355	9·0	310	4·0	350	7·5	30	5·0	355	5·0	345	Surface
0·5	10·0	345	13·0	315	5·5	360	14·0	35	9·0	10	8·5	355	0·5
1·0	15·5	345	17·0	325	6·0	355	16·0	65	9·5	10	16·0	350	1·0
1·5	18·0	345	20·0	335	7·5	350	14·5	60	8·5	15	14·5	320	1·5
2·0	19·0	345	22·0	330			15·0	50	9·5	20	10·0	325	2·0
2·5	22·0	345	25·0	330			13·5	30	10·5	15	16·0	330	2·5
3·0	25·0	345	29·0	325			16·0	35	12·0	5	18·5	335	3·0
3·5			30·0	325			17·5	30	13·0	30	16·0	345	3·5
4·0									15·0	25	9·5	350	4·0
4·5									19·0	25	13·5	325	4·5
5·0											25·0	325	5·0
	2 theodolites from 1·3 km. to 2·6 km.		2 theodolites to 2·4 km.		2 theodolites burst ?		distance		distance		distance		

Class (b)—continued.

	April 19, 1909 8.15 a.m.		April 26, 1909 2.41 p.m.		May 4, 1909 8.15 a.m.		May 31, 1909 0.13 p.m.		June 21, 1909 7.0 p.m.		
	vel.	dir.	vel.	dir.	vel.	dir.	vel.	dir.	vel.	dir.	
Gradient	7·0	170	8·0	185	7·0	160	7·0	225	8·0	200	Gradient
Surface	4·0	120	5·5	160	5·5	135	5·0	160	5·5	165	Surface
0·5	8·5	135	10·0	160	10·0	140	10·0	175	6·5	175	0·5
1·0	7·0	155	10·5	200	13·0	125	9·0	185	9·5	215	1·0
1·5	7·5	185	16·0	220	13·0	125	4·5	200	12·5	210	1·5
2·0	7·0	185	16·5	220	12·5	125	7·5	235	14·0	210	2·0
2·5	8·5	195	14·0	215	5·5	120	10·0	245	13·0	200	2·5
3·0	8·0	205	[15·5	215]	9·0	125	11·0	230	14·0	200	3·0
3·5	6·0	225					13·5	225	16·0	195	3·5
4·0	7·0	205					12·0	230	22·5	205	4·0
4·5	11·0	200									4·5
					burst		theodolite accidentally moved				

Class (c).

	Feb. 26, 1907 noon		April 4, 1907 6.34 p.m.		April 16, 1907 6.27 p.m.		May 13, 1907 6.27 p.m.		May 29, 1907 7.11 p.m.		Dec. 26, 1907 0.4 p.m.		
	vel.	dir.	vel.	dir.	vel.	dir.	vel.	dir.	vel.	dir.	vel.	dir.	
Gradient	—	—	4·5	90	6·0	50	9·0	345	11·0	140	11·0	135	Gradient
Surface	2·5	30	4·5	80	3·0	360	4·5	320	3·0	70	7·5	80	Surface
0·5	6·5	35	6·5	80	8·5	10	11·0	340	8·5	70	10·0	100	0·5
1·0	8·0	45	6·0	70	5·0	35	0·5	360	4·5	125	17·5	110	1·0
1·5	[4·5	45]	5·5	65	4·5	25			1·5	80	10·0	115	1·5
2·0			4·0	70							5·0	90	2·0
2·5			5·0	50							3·5	70	2·5
3·0			4·0	360							2·5	355	3·0
3·5			3·5	5							6·0	55	3·5
4·0											5·0	60	4·0
	cloud				cloud		Totland Bay cloud		Totland Bay cloud		distance		

CLASS (c)—continued.

Gradient	Jan. 2, 1908 10.43 a.m. vel.	dir.	Jan. 2, 1908 3.47 p.m. vel.	dir.	Jan. 3, 1908 10.49 a.m. vel.	dir.	Jan. 3, 1908 11.19 a.m. vel.	dir.	Jan. 3, 1908 3.59 p.m. vel.	dir.	Jan. 4, 1908 11.2 a.m. vel.	dir.	Gradient
Gradient	19·0	105	16·0	105	20·0	105	20·0	105	16·5	100	19·0	90	Gradient
Surface	9·5	55	10·0	50	7·0	55	8·5	55	9·0	50	7·5	70	Surface
0·5	10·0	65	16·0	60	11·0	70	17·5	65	16·5	60	15·0	75	0·5
1·0	22·5	80	14·0	80	16·0	85	20·0	85	21·5	85	14·0	95	1·0
1·5	7·0	85	12·0	90	11·5	115	20·0	105	13·0	105	14·0	100	1·5
2·0	10·0	65	10·0	90			8·0	120	11·0	100			2·0
2·5	9·0	50					10·0	95	10·0	80			2·5
3·0	8·0	35							7·5	110			3·0
3·5	10·5	40							6·5	105			3·5
4·0	10·5	30							8·0	100			4·0
4·5									9·5	100			4·5
5·0									11·0	90			5·0
	distance				lost behind tree						lost behind tree		

Gradient	Mar. 14, 1908 10.37 a.m. vel.	dir.	Mar. 27, 1908 5.41 p.m. vel.	dir.	April 9, 1908 11.8 a.m. vel.	dir.	April 16, 1908 11.12 a.m. vel.	dir.	May 29, 1908 8.12 a.m. vel.	dir.	June 4, 1908 7.0 p.m. vel.	dir.	Gradient
Gradient	8·0?	115?	13·0	230	—	—	25·0	90	20·0	70	9·0	360	Gradient
Surface	6·0	100	5·0	205	2·0	25	17·0	55	9·0	40	12·5	25	Surface
0·5	5·5	125	8·5	215	4·0	65	18·5	65	12·0	50	14·5	25	0·5
1·0	2·0	115	11·5	235	7·5	10	20·5	80	12·5	70	6·5	20	1·0
1·5	0·5	65	10·0	250	5·0	5	20·0	90	14·5	80	3·5	330	1·5
2·0			8·5	250	5·0	350	14·0	90	11·5	80			2·0
2·5			7·5	260	5·0	25			13·0	65			2·5
3·0			7·0	280	3·5	355							3·0
3·5					4·0	325							3·5
4·0					[3·5	315]							4·0
	2 theodolites burst				2 theodolites to 0·7 km.		2 theodolites to 2·0 km.						

Class (c)—continued.

	Jan. 20, 1909 10.20 a.m.		Feb. 12, 1909 5.8 p.m.		Feb. 19, 1909 10.2 a.m.		Feb. 21, 1909 2.35 p.m.		Feb. 22, 1909 11.38 a.m.		May 4, 1909 7.4 p.m.		
	vel.	dir.	vel.	dir.	vel.	dir.	vel.	dir.	vel.	dir.	vel.	dir.	
Gradient	7·0?	30?	23·0	60	11·0	145	10·5	155	—	—	13·0	150	Gradient
Surface	7·5	5	18·0	30	7·0	100	4·0	115	5·0	80	7·0	80	Surface
0·5	12·5	30	19·0	30	11·0	115	10·0	145	9·0	105	13·0	90	0·5
1·0	11·0	15	24·0	50	15·5	130	11·0	155	7·5	110	10·5	120	1·0
1·5	12·5	10	27·5	70	15·0	130	7·5	155	8·0	110	11·5	140	1·5
2·0	14·0	25	24·0	75	13·0	130	7·0	140	7·5	110	12·0	135	2·0
2·5	12·0	40	22·5	60	8·5	140	5·0	110	6·0	70	11·0	130	2·5
3·0	[7·0	45]	11·5	65			6·0	120	4·0	40	7·0	105	3·0
3·5			12·0	45			6·5	130			8·0	100	3·5
4·0			17·5	55			6·0	170			9·0	105	4·0
4·5							3·5	185			7·5	90	4·5
5·0							5·0	170			7·0	100	5·0
5·5							5·0	175			5·0	115	5·5
			distance		2 theodolites haze		distance						

	May 5, 1909 8.10 a.m.		May 6, 1909 10.33 a.m.		June 3, 1909 0.5 p.m.		
	vel.	dir.	vel.	dir.	vel.	dir.	
Gradient	14·0	130	24·5?	120	22·0	80	Gradient
Surface	5·5	90	4·5	85	10·0	40	Surface
0·5	12·0	100	8·0	95	14·0	45	0·5
1·0	11·0	85	18·5	105	12·0	95	1·0
1·5	12·0	90	13·5	115			1·5
2·0	11·0	100	18·5	115			2·0
2·5			10·0	130			2·5
	burst?		distance		cloud		

CLASS (d). Reversals or great changes in direction.

	Jan. 25, 1907 11.42 a.m.		Mar. 30, 1907 6.19 p.m.		April 1, 1907 6.12 p.m.		April 9, 1907 6.20 p.m.		April 15, 1907 10.50 a.m.		April 15, 1907 6.86 p.m.		
	vel.	dir.	vel.	dir.	vel.	dir.	vel.	dir.	vel.	dir.	vel.	dir.	
Gradient	—	—	—	—	3·5	180	—	—	9·0	70	9·0	360	Gradient
Surface	6·0?	25	4·5	255	3·0	185	2·5	170	8·5	35	6·0	10	Surface
0·5	6·0	10	4·0	250	3·0	210	2·5	170	9·0	40	7·0	20	0·5
1·0	1·0	310	2·0	325	3·5	150	0·5	35	2·0	205	2·0	100	1·0
1·5	3·0	230	[3·5	55]	3·5	115	1·0	190	2·0	235	[2·0	70]	1·5
2·0					3·5	130							2·0
2·5					4·0	150							2·5
3·0					1·5	220							3·0
3·5					4·0	210							3·5
4·0					[5·0	190]							4·0

at 0·3 km. direction was 55. altitude became too great for the theodolite used — veering began at 0·7 km. — a veer of 75° between 2·8 and 3 km. — between 0·7 and 1·6 km. the wind dir. went from 190 through E., N. and S. to 165. cloud — sudden change of direction at 0·8 km. accompanied by decreased velocity. cloud — wind veered to 145° at 1·2 km. then changed suddenly to 70°. cloud

	May 14, 1907 6.53 p.m.		May 17, 1907 6.43 p.m.		May 21, 1907 6.8 p.m.		May 25, 1907 7.13 p.m.		May 27, 1907 6.11 p.m.		June 29, 1907 7.11 p.m.		
	vel.	dir.	vel.	dir.	vel.	dir.	vel.	dir.	vel.	dir.	vel.	dir.	
Gradient	—	—	7·0	10	5·0	90	6·0	95	—	—	—	—	Gradient
Surface	1·5	30	4·5	305	1·0	135	5·0	50	4·0	180	6·0	340	Surface
0·5	4·0	40	7·5	310	4·0	90	10·0	95	2·5	350	11·0	330	0·5
1·0	1·0	325	7·5	310	3·5	140	7·5	140	5·5	20	4·0	275	1·0
1·5	7·0	250	7·5	255	1·5	215	5·5	180	6·0	15	1·5	275	1·5
2·0	8·0	245	[8·0	240]	2·5	265	6·0	205	2·5	360	0·5	200	2·0
2·5					10·0	230	6·5	185	3·5	310	2·5	180	2·5
3·0									1·0	205			3·0
3·5									1·5	225			3·5
4·0									3·5	180			4·0
4·5									8·0	225			4·5
5·0									7·5	220			5·0
5·5									8·0	220			5·5

Continued on next page.

CLASS (d)—continued.

	May 14, 1907 6.53 p.m.	May 17, 1907 6.43 p.m.	May 21, 1907 6.8 p.m.	May 25, 1907 7.13 p.m.	May 27, 1907 6.11 p.m.		June 29, 1907 7.11 p.m.	
					vel.	dir.		
6·0					7·5	230		6·0
6·5					6·0	240		6·5
	Totland Bay cloud	Totland Bay	Totland Bay wind backed to 75° at 0·7 km. cloud	Totland Bay cloud	Totland Bay wind veered between surface and 0·5 km. haze			

	July 1, 1907 3.15 p.m.		July 1, 1907 4.7 p.m.		July 22, 1907 0.47 p.m.		Aug. 5, 1907 2.24 p.m.		Aug. 23, 1907 7.5 p.m.		Aug. 28, 1907 11.38 a.m.		
	vel.	dir.	vel.	dir.	vel.	dir.	vel.	dir.	vel.	dir.	vel.	dir.	
Gradient	8·0?	15?	—	—	—	—	6·0	310	7·0	330	—	—	Gradient
Surface	5·5	40	4·5	20	6·0	175	1·0	60	3·0	155	4·5	115	Surface
0·5	4·0	40	6·0	10	4·0	175	1·0	360	7·0	160	4·5	125	0·5
1·0	2·5	25	6·0	25	3·0	240	4·0	240	6·0	185	6·0	145	1·0
1·5			7·0	10	6·0	265			3·0	155	4·5	130	1·5
2·0			6·0	350					9·5	170	4·0	135	2·0
2·5									8·5	130	6·5	235	2·5
3·0									6·0	140			3·0
3·5									4·0	120?			3·5
	Chobham Common reversal above. 2 theodolites cloud		Chobham Common reversal above. cloud		2 theodolites				reversal above				

	Aug. 28, 1907 6.23 p.m.		Oct. 14, 1907 5.24 p.m.		Mar. 3, 1908 10.12 a.m.		Mar. 12, 1908 5.52 p.m.		Mar. 13, 1908 5.40 p.m.		Mar. 21, 1908 5.50 p.m.		
	vel.	dir.	vel.	dir.	vel.	dir.	vel.	dir.	vel.	dir.	vel.	dir.	
Gradient	—	—	18·0	205	7·0?	—	—	—	2·0	50	6·0	190	Gradient
Surface	1·5	165	7·5	340	2·0	170	1·5	215	3·0	75	5·5	135	Surface
0·5	2·0	100	10·0	330	2·5	210	2·0	260	2·5	50	4·0	130	0·5
1·0	5·0	240	10·5	180	4·0	245	3·0	340	3·0	350	4·0	165	1·0
1·5	6·0	225			5·5	225	6·0	335	2·5	300	4·0	110	1·5
2·0	8·0	235			5·0	220	5·5	350	3·0	315	5·0	200	2·0
2·5	9·0	235			3·5	215	5·0	350			5·5	200	2·5

Continued on next page.

Class (d)—continued.

	Aug. 28, 1907 6.23 p.m.		Oct. 14, 1907 5.24 p.m.	Mar. 3, 1908 10.12 a.m.		Mar. 12, 1908 5.52 p.m.		Mar. 13, 1908 5.40 p.m.	Mar. 21, 1908 5.50 p.m.		
	vel.	dir.		vel.	dir.	vel.	dir.		vel.	dir.	
3·0	8·0	240		5·0	245	3·5	350		8·5	200	3·0
3·5	9·5	250		4·5	225				3·0	245	3·5
4·0	6·0	255							7·0	195	4·0
4·5	7·0	235							7·5	210	4·5
5·0	6·0	230							4·5	215	5·0
5·5	6·0	235									5·5
6·0	6·5	235									6·0
	backing of 285° between surface and 1·0 km.		wind veered between 0·5 and 1·0 km. with very small velocity	at about 25 km. sudden veer from 215° to 295°. haze		2 theodolites burst		2 theodolites cloud	2 theodolites to 4·5 km.		

	April 9, 1908 5.53 p.m.		June 2, 1908 7.15 p.m.		July 27, 1908 7.19 p.m.		Sept. 30, 1908 8.36 a.m.		Sept. 30, 1908 11.17 a.m.		Nov. 3, 1908 0.47 p.m.		
	vel.	dir.	vel.	dir.	vel.	dir.	vel.	dir.	vel.	dir.	vel.	dir.	
Gradient	—	—	5·0	160	3·0	300	7·0	210	8·0	220	8·0	120	Gradient
Surface	2·0	205	1·0	145	4·0	260	3·5	175	5·0	160	2·5	90	Surface
0·5	3·0	350	2·5	190	4·0	270	6·5	180	8·5	175	4·0	130	0·5
1·0	6·0	330	5·5	220	2·0	330?	2·5	180	11·0	185	4·5	105	1·0
1·5	9·0	355	6·5	220	2·0	210	0·5	330	4·0	150	4·0	60	1·5
2·0	6·5	360	7·5	205	7·0	200	3·5	240	2·0	90	5·0	50	2·0
2·5	4·0	320	6·5	210	7·0	205	3·5	170	0·5	50	4·0	15	2·5
3·0	5·0	300	6·5	185	6·5	220	4·0	235	4·5	230	3·0	10	3·0
3·5	4·5	295	8·5	190	6·0	215	6·0	230	7·0	185	4·0	345	3·5
4·0	8·5	305	9·5	195	7·5	215			7·0	215	3·0	320	4·0
4·5	9·0	290	9·5	200	7·5	205			7·0	230	4·5	300	4·5
5·0			10·0	210	12·0	210			7·5	260	5·5	320	5·0
5·5			10·0	210	13·0	210			12·0	220	5·5	330	5·5
6·0			9·5	295	18·0	220			16·0	225	6·5	340	6·0
6·5					21·5	220			16·5	215	9·0	320	6·5
7·0					22·0	215							7·0

Continued on next page.

C.

13

CLASS (d)—continued.

	April 9, 1908 5.53 p.m.	June 2, 1908 7.15 p.m.	July 27, 1908 7.19 p.m.	Sept. 30, 1908 8.36 a.m.	Sept. 30, 1908 11.17 a.m.	Nov. 3, 1908 0.47 p.m.	
			vel.　dir.				
7·5			23·0　210				7·5
8·0			25·0　215				8·0
8·5			25·0　215				8·5
			2 theodolites to 1·4 km. veer of 195° between 0·9 and 1·2 km.				

	Nov. 6, 1908 10.59 a.m.		Nov. 7, 1908 3.25 p.m.		Nov. 8, 1908 4.24 p.m.		June 1, 1909 10.28 a.m.		
	vel.	dir.	vel.	dir.	vel.	dir.	vel.	dir.	
Gradient	11·0	100	18·0	110	23·0	100	—	—	Gradient
Surface	6·0	100	7·5	80	9·0	45	9·0	20	Surface
0·5	11·0	110	11·0	85	17·5	55	14·0	25	0·5
1·0	16·0	120	9·5	105	25·0	80	5·5	105	1·0
1·5	8·0	125	6·5	140	20·0	80	8·5	150	1·5
2·0	2·0	100	2·0	110	17·0	85	9·5	175	2·0
2·5	3·5	125	4·0	210	13·0	90	[9·0	190]	2·5
3·0	1·5	120	3·0	225	8·5	75			3·0
3·5	1·0	360	2·0	235	4·5	30			3·5
4·0	1·5	280	4·0	240	4·0	340			4·0
4·5	4·0	300	4·5	230	3·0	300			4·5
5·0	5·5	305	6·5	240					5·0
5·5	6·0	290	5·5	265					5·5
6·0	7·0	290	8·0	260					6·0
6·5	8·0	290							6·5
7·0	8·5	300							7·0
7·5	8·5	305							7·5
8·0	9·0	300							8·0
8·5	10·5	300							8·5
9·0	12·5	295							9·0
9·5	14·5	300							9·5
	burst		maximum velocity 14 at 0·7 km.				cloud		

CLASS (e).

Upper wind blowing westwards from centres of low pressure; frequently reversals at a lower layer.

1. Upper wind between West and North.

	Feb. 8, 1907 0.45 p.m.		April 18, 1907 6.40 p.m.		April 19, 1907 5.59 p.m.		April 20, 1907 4.52 p.m.		Sept. 4, 1907 5.4 p.m.		Sept. 23, 1907 2.17 p.m.		
	vel.	dir.	vel.	dir.	vel.	dir.	vel.	dir.	vel.	dir.	vel.	dir.	
Gradient	8·0	205	5·0	330	—	—	16·0	225	8·0	230	2·0	110	Gradient
Surface	2·5	180	2·5	30	3·0	190	5·5	190	7·0	220	1·0	115	Surface
0·5	6·5	175	3·0	15	5·5	230	12·0	210	7·0	220	2·0	110	0·5
1·0	[3·5	300]	3·0	350	3·5	340	12·0	220	5·0	175	3·0	120	1·0
1·5			2·5	355	5·0	5	10·0	250	5·5	275	2·0	145	1·5
2·0			5·0	350	6·0	360	7·0	265	10·5	280	1·5	180	2·0
2·5			6·5	340	5·0	5	7·0	270			2·0	210	2·5
3·0			8·0	330	6·5	350	8·0	295			1·5	235	3·0
3·5			9·5	330			8·5	305			3·5	280	3·5
4·0			14·5	330			8·5	325			4·0	270	4·0
4·5			17·5	335							1·5	240	4·5
5·0			16·5	325							4·5	290	5·0
5·5											3·5	295	5·5
6·0											2·5	280	6·0
6·5											2·0	200	6·5
7·0											1·0	295	7·0
7·5											0	—	7·5
8·0											2·0	225	8·0
									2 theodolites to 1 km. burst		2 theodolites to 3 km.		

	Jan. 11, 1908 3.38 p.m.		Feb. 14, 1908 4.38 p.m.		Mar. 5, 1908 11.30 a.m.		Mar. 5, 1908 5.5 p.m.		Mar. 24, 1908 10.19 a.m.		Mar. 27, 1908 11.4 a.m.		
	vel.	dir.	vel.	dir.	vel.	dir.	vel.	dir.	vel.	dir.	vel.	dir.	
Gradient	8·0	240	12·0	280	7·5?	250?	9·0	220	—	—	—	—	Gradient
Surface	3·0	170	7·5	280	—	—	—	—	4·5	185	4·5	200	Surface
0·5	4·5	180	9·5	285	5·0	235	6·0	210	8·0	200	4	190	0·5
1·0	1·5	230	12·5	300	5·0	240	8·0	220	7·5	205	6	210	1·0

Continued on next page.

CLASS (e)—continued.

	Jan. 11, 1908 3.38 p.m.		Feb. 14, 1908 4.38 p.m.		Mar. 5, 1908 11.30 a.m.		Mar. 5, 1908 5.5 p.m.		Mar. 24, 1908 10.19 a.m.		Mar. 27, 1908 11.4 a.m.		
	vel.	dir.	vel.	dir.	vel.	dir.	vel.	dir.	vel.	dir.	vel.	dir.	
1·5	2·5	230	10·0	280	3·0	210	8·0	240	9·0	210	3	240	1·5
2·0	4·0	350	10·5	280	5·0	240	12·0	240	9·5	220	3	265	2·0
2·5	3·5	310	12·0	275	6·5	245			3·5	260	4·5	335	2·5
3·0	3·5	335	10·5	270	9·5	260			5·0	5			3·0
3·5	5·0	350	8·5	270	[12·0	280]			10·5	360			3·5
4·0	2·0	60	8·5	275					9·0	330			4·0
4·5	6·0	105	10·0	290					10·5	340			4·5
5·0	7·0	80	12·5	335									5·0
5·5	5·0	70											5·5
6·0	6·0	30											6·0
			2 theodolites to 1·8 km.		2 theodolites to 1·5 km. haze		2 theodolites burst		2 theodolites to 0·3 km. burst		burst		

	Mar. 28, 1908 5.40 p.m.		April 1, 1908 5.50 p.m.		April 1, 1908 6.13 p.m.		April 29, 1908 3.57 p.m.		June 10, 1908 7.15 p.m.		June 22, 1908 6.36 p.m.		
	vel.	dir.	vel.	dir.	vel.	dir.	vel.	dir.	vel.	dir.	vel.	dir.	
Gradient	10·0	270	14·0	280	14·0	280	—	—	8·0	270	5·0	310	Gradient
Surface	6·0	295	6·0	295	7·0	295	5·0	220	4·0	270	5·5	240	Surface
0·5	6·0	300	7·0	280	7·5	285	6·0	235	8·0	260	4·5	290	0·5
1·0	6·0	305	8·5	300	9·5	310	5·0	240	4·0	250	5·5	340	1·0
1·5	4·5	325	14·0	315			4·0	250	2·5	220	3·0	25	1·5
2·0	10·0	345	21·5	320			8·0	250	7·0	250	4·5	5	2·0
2·5	10·0	330					6·0	250	8·0	265	5·0	350	2·5
3·0	12·0	330					6·5	265	8·5	275	4·5	360	3·0
3·5							6·5	290	8·0	275	6·0	35	3·5
4·0							8·5	280	8·5	295	6·0	15	4·0
4·5							10·5	285	8·0	295	6·0	20	4·5
5·0							15·0	300	8·0	300	6·5	25	5·0
5·5							17·0	300	8·0	310	6·0	25	5·5
6·0							20·5	300	8·0	320	6·0	30	6·0

Continued on next page.

CLASS (e)—continued.

	Mar. 28, 1908 5.40 p.m.	April 1, 1908 5.50 p.m.	April 1, 1908 6.13 p.m.	April 29, 1908 3.57 p.m.	June 10, 1908 7.15 p.m.	June 22, 1908 6.36 p.m.		
						vel.	dir.	
6·5						6·0	25	6·5
7·0						6·0	30	7·0
7·5						6·0	45	7·5
8·0						5·0	40	8·0
8·5						7·0	50	8·5
9·0						8·0	50	9·0
9·5						7·5	40	9·5
10·0						7·5	55	10·0
10·5						8·0	55	10·5
11·0						6·0	50	11·0
11·5						5·5	55	11·5
12·0						5·5	60	12·0
12·5						5·5	15	12·5
13·0						8·0	30	13·0
	2 theodolites burst	2 theodolites cloud?	2 theodolites cloud	cloud		2 theodolites to 4·3 km. burst		

	July 30, 1908 8.6 a.m.		July 30, 1908 0.36 p.m.		Aug. 1, 1908 8.10 a.m.		Aug. 1, 1908 5.53 p.m.		Aug. 1, 1908 7.5 p.m.		Aug. 2, 1908 5.54 p.m.		
	vel.	dir.	vel.	dir.	vel.	dir.	vel.	dir.	vel.	dir.	vel.	dir.	
Gradient	—	—	6·0	330	—	—	5·0	20	5·0	20	—	—	Gradient
Surface	2·5	270	3·0	300	5·0	355	4·0	350	6·5	345	4·0	175	Surface
0·5	4·5	280	3·0	285	7·0	5	5·0	350	9·0	355	2·5	360	0·5
1·0	1·5	275	3·0	285	6·0	355	4·5	345	5·0	10	7·0	320	1·0
1·5	5·0	330	3·0	320	12·0	340	10·5	5	9·0	5	10·5	320	1·5
2·0	5·0	350	4·0	330	11·0	340	13·0	355	12·5	360	13·5	330	2·0
2·5	7·0	10	5·5	320	11·0	345	14·5	5	13·0	360	13·5	325	2·5
3·0	8·5	360	7·0	320	13·0	350	13·5	355	14·5	350	13·0	320	3·0
3·5	7·5	360	8·0	320	[16·0	345]	15·5	350	16·5	355	14·0	320	3·5
4·0	7·5	345	5·0	310			18·5	360	18·5	360	17·0	310	4·0
4·5	5·0	350											4·5
5·0	8·0	350											5·0
5·5	[8·0	355]											5·5

CLASS (e)—continued.

Gradient	Aug. 2, 1908 7.30 p.m. vel.	dir.	Nov. 16, 1908 10.47 a.m. vel.	dir.	Feb. 7, 1909 4.28 p.m. vel.	dir.	Feb. 17, 1909 8.17 a.m. vel.	dir.	Feb. 18, 1909 4.43 p.m. vel.	dir.	Mar. 5, 1909 5.13 p.m. vel.	dir.	May 2, 1909 7.7 p.m. vel.	dir.	Gradient
Gradient	—	—	—	—	11·0	185	—	—	13·0	120	15·0	200	—	—	Gradient
Surface	1·0	125	5·0	340	7·5	135	2·5	290	4·0	115	3·5	160	2·5	250	Surface
0·5	4·0	45	5·0	335	11·0	140	3·5	290	8·0	115	9·0	180	4·0	265	0·5
1·0	6·5	20	4·0	295	11·5	130	4·0	290	10·5	110	7·0	200	2·5	350	1·0
1·5	9·0	30	5·0	265	6·0	155	3·0	80	7·0	130	7·5	210	2·0	10	1·5
2·0	13·0	40	3·0	160	6·0	150	2·5	55	6·0	130	3·5	220	4·0	10	2·0
2·5	13·0	30	2·5	—	2·5	—	3·5	340	3·0	125	5·0	250	9·0	345	2·5
3·0	12·5	25	5·5	5	3·0	270	8·0	320	1·0	140	8·0	260	9·5	350	3·0
3·5	13·0	25	4·0	355	2·5	320	10·0	320	2·0	110	9·0	270	11·5	350	3·5
4·0	13·5	10	4·5	330	2·5	290	10·5	335	2·0	360	12·0	270	12·0	340	4·0
4·5	14·5	10	4·0	10			9·5	310	3·0	30	15·0	275	14·5	350	4·5
5·0	[18·0	10]	7·5	10			7·5	310	2·0	335			20·0	335	5·0
5·5			[10·5	5]			7·0	305	1·5	330?					5·5
6·0							7·5	295	3·0	310					6·0
6·5									5·0	305					6·5
7·0									6·0	320					7·0
7·5									8·5	325					7·5
8·0									[10·0	335]					8·0
			2 theodolites to 2·6 km.		burst				2 theodolites to 5·6 km.						

Class (e)—continued.

2. Upper wind between South and West.

Gradient	July 24, 1907 8.7 p.m.		May 2, 1908 0.48 p.m.		June 3, 1908 10.19 a.m.		June 3, 1908 6.54 p.m.		May 16, 1909 6.22 p.m.		June 3, 1909 0.59 p.m.		Gradient
	vel.	dir.	vel.	dir.	vel.	dir.	vel.	dir.	vel.	dir.	vel.	dir.	
Gradient	6·0	125	—	—	6·0	145	5·0	40	9·0	70	22·0	80	Gradient
Surface	4·0	140	4·5	190	2·0	60	6·5	50	4·5	50	15·0	35	Surface
0·5	4·5	120	4·0	165	4·0	60	6·0	80	9·0	50	16·0	45	0·5
1·0	2·0	50	2·0	120	5·5	80	4·5	110	8·5	70	10·0	95	1·0
1·5	2·0	150	2·0	110	6·0	105	3·0	110	4·0	120	7·5	110	1·5
2·0	1·5	195	2·5	120	5·0	110	4·0	160	6·0	150	7·5	100	2·0
2·5	2·0	220	1·5	120	2·5	130	6·0	155	5·5	190	9·0	105	2·5
3·0	3·0	215	2·0	125	2·5	170	5·0	150	6·5	185	8·0	120	3·0
3·5	3·0	220	1·5	185	6·0	180	5·0	160			7·0	140	3·5
4·0			3·5	270	9·0	175	5·0	165					4·0
4·5			6·0	270	8·5	180	5·0	160					4·5
5·0			6·0	290	[11·0	165]	5·5	180					5·0
5·5			7·0	250			4·0	195					5·5
6·0			9·0	240			5·0	170					6·0
6·5			10·0	245			3·5	170					6·5
7·0							7·5	150					7·0
7·5							8·5	155					7·5
8·0							6·5	170					8·0
8·5							4·0	170					8·5
9·0							4·0	250					9·0
9·5							4·5	230					9·5
10·0							4·0	200					10·0
10·5							2·0	170					10·5
11·0							5·0	180					11·0
11·5							9·0	205					11·5
	2 theodolites to 0·8 km.		2 theodolites to 4·0 km.		2 theodolites to 2·8 km.		2 theodolites to 9·0 km. distance		burst				

CLASS (*f*). The Stratosphere.

Gradient	July 28, 1908 7.0 p.m.		July 29, 1908 7.0 p.m.		July 30, 1908 7.0 p.m.		July 31, 1908 7.0 p.m.		Sept. 30, 1908 4.31 p.m.		Oct. 1, 1908 4.20 p.m.		Gradient
	vel.	dir.	vel.	dir.	vel.	dir.	vel.	dir.	vel.	dir.	vel.	dir.	
Gradient	9·0	40	9·0	40	7·0	320	—	—	6·0	180	4·0	160	Gradient
Surface	—	35	5·0	360	—	—	5·0	355	4·5	150	3·0	150	Surface
0·5	—	—	5·0	360	6·5	280	5·0	355	4·5	175	3·0	150	0·5
1·0	—	—	4·5	10	8·0	285	5·0	360	4·0	150	4·5	155	1·0
1·5	—	—	4·0	5	8·5	290	4·0	360	3·5	160	7·0	175	1·5
2·0	2·5	330	5·0	30	6·0	285	6·0	340	3·5	185	8·5	175	2·0
2·5	5·5	330	5·5	25	8·0	270	6·5	330	3·0	175	7·5	175	2·5
3·0	6·5	330	6·0	15	10·5	275	10·0	330	2·5	180	10·0	185	3·0
3·5	6·5	340	6·5	360	13·5	290	19·0	325	5·0	220	11·0	180	3·5
4·0	7·0	335	7·5	5	13·5	305	19·0	320	10·0	230	13·5	175	4·0
4·5	8·0	340	10·0	10	9·5	300	21·5	315	9·0	210	14·0	175	4·5
5·0	11·5	350	11·5	5	9·5	305	21·5	310	9·5	210	13·0	185	5·0
5·5	15·0	350	10·0	5	12·0	310	20·5	310	10·0	215	12·5	195	5·5
6·0	15·0	355	8·0	5	12·5	310	21·5	310	9·5	220	12·0	210	6·0
6·5	15·0	350	8·0	5	12·5	305	22·5	310	10·5	215	16·0	215	6·5
7·0	16·5	350	10·0	5	12·5	295	20·5	305	12·5	205	15·5	210	7·0
7·5	20·0	350	12·5	20	11·5	300	19·0	310	11·5	205	15·5	210	7·5
8·0	23·0	360	12·5	20	10·5	300	24·5	320	12·0	215	17·0	210	8·0
8·5	23·0	360	14·5	15	11·0	300	27·5	320	12·0	220	19·5	210	8·5
9·0	23·5	350	15·0	10	13·0	310	30·5	320	14·0	220	21·0	210	9·0
9·5	22·0	355	15·0	10	12·0	320	29·5	310	16·5	220	22·0	215	9·5
10·0	22·0	350	18·5	10	12·0	305	29·5	310	17·5	230	23·5	205	10·0
10·5	20·5	340	19·0	5	14·5	305	31·5	300	20·0	230	24·0	210	10·5
11·0	20·5	340	20·0	5	15·0	315	33·0	305	20·0	230	24·0	210	11·0
11·5	23·0	340	24·0	5	13·0	320	35·0	300	23·0	225	25·0	195	11·5
12·0	16·5	340	22·0	5			33·0	300	25·5	225	26·0	195	12·0
12·5	9·5	340	18·0	10			30·0	305	25·0	225	26·0	190	12·5
13·0	4·0	310	13·0	350			27·0	315	25·5	230	23·0	190	13·0

Continued on next page.

CLASS (f)—continued.

	July 28, 1908 7.0 p.m.	July 29, 1908 7.0 p.m.	July 30, 1908 7.0 p.m.	July 31, 1908 7.0 p.m.		Sept. 30, 1908 4.31 p.m.		Oct. 1, 1908 4.20 p.m.		
				vel.	dir.	vel.	dir.	vel.	dir.	
13·5				24·0	310	28·5	230	19 0	190	13·5
14·0						27·0	230	14·0	205	14·0
14·5						23·5	235	10·5	210	14·5
15·0						19·0	225	9·0	210	15·0
15·5						13·0	205	7·0	175	15·5
16·0						13·0	199	7·0	180	16·0
16·5								9·0	170	16·5
17·0								8·5	180	17·0
17·5								7·0	180	17·5
	2 theodolites to 7·7 km. distance	2 theodolites to 13 km. burst	2 theodolites to 9 km. distance	2 theodolites to 10 km. burst		2 theodolites to 3·5 km. distance		2 theodolites to 4 km. burst		

	Oct. 2, 1908 4.20 p.m.		May 6, 1909 6.25 p.m.		May 7, 1909 2.52 p.m.		May 7, 1909 6.29 p.m.		Aug. 5, 1909 6.33 p.m.		Mar. 3, 1910 4.30 p.m.		
	vel.	dir.	vel.	dir.	vel.	dir.	vel.	dir.	vel.	dir.	vel.	dir.	
Gradient	6 0	160	21·0	120	22·0	115	22·0	115	4·0	90	—	—	Gradient
Surface	2·5	130	—	—	15·0	—	12·2	70	—	—	8·5	115	Surface
0·5	5·5	125	—	—	—	—	15·0	70	2·0	45	9·5	105	0·5
1·0	4·0	140	12·0	100	—	—	18·0	95	4·5	70	10·0	125	1·0
1·5	7·0	160	12·0	110	14·0	100	14·0	95	5·5	80	11·0	135	1·5
2·0	8·5	160	11·0	115	10·0	105	9·0	100	5·0	40	12·0	140	2·0
2·5	9·0	160	10·0	115	11·0	100	9·0	100	5·0	80	9·0	145	2·5
3·0	12·5	155	10·0	120	10·5	100	13·5	100	6·0	85	8·0	145	3·0
3·5	9·5	165	9·5	120	10·0	95	9·0	90	3·5	65	4·5	90	3·5
4·0	7·0	155	9·5	90	14·0	100	10·5	90	3·0	60	7·0	95	4·0
4·5	8·5	150	10·0	110	9·0	100	15·0	90	4·0	55	5·0	50	4·5
5·0	10·0	155	10·0	120	8·0	115	11·5	90	5·0	60	6·0	70	5·0
5·5	12·0	170	12·0	130	7·0	120	6·5	85	4·0	45	5·5	65	5·5
6·0	10·5	180	12·5	140	7·5	100	4·5	80	2·5	25	5·5	55	6·0
6·5	8·0	185	11·0	135	6·0	80	5·0	65	2·0	40	4·5	45	6·5
7·0	8·5	180	11·0	145	4·5	65	7·5	60	3·0	70	4·0	40	7·0

Continued on next page.

C.   14

CLASS (f)—continued.

	Oct. 2, 1908 4.20 p.m.		May 6, 1909 6.25 p.m.		May 7, 1909 2.52 p.m.		May 7, 1909 6.29 p.m.		Aug. 5, 1909 6.33 p.m.		Mar. 3, 1910 4.30 p.m.		
	vel.	dir.	vel.	dir.	vel.	dir.	vel.	dir.	vel.	dir.	vel.	dir.	
7·5	7·0	190	9·0	155	7·5	80	7·5	80	3·0	80	3·5	25	7·5
8·0	9·5	190	13·0	140	8·5	105	6·5	115	0·0	—	5·5	25	8·0
8·5	10·5	200	16·0	135	9·0	120	2·0	100	5·0	30	5·0	20	8·5
9·0	10·0	215	17·0	130	7·5	125	5·5	110	9·0	25	5·0	10	9·0
9·5	9·5	225	16·0	130	6·0	155	4·5	100	5·0	30	5·0	15	9·5
10·0	11·0	215	18·0	130	5·5	155	3·0	100	5·0	40	6·5	45	10·0
10·5	11·0	215	20·0	130	7·5	140	3·5	135	4·0	30	5·5	30	10·5
11·0	12·0	215	20·0	135	9·0	130	4·5	140	3·5	35	4·0	20	11·0
11·5	12·0	220	15·0	145	6·0	145	6·0	125	3·5	40	5·0	80	11·5
12·0	12·0	225	14·0	140	5·5	130	4 5	—	1·0	—	1·0	190	12·0
12·5	12·5	220	12·0	115	2·5	165	4·0	—	2·5	350	5·0	265	12·5
13·0	12·5	220	[8·0	145]	2·0	210	2·0	—	1·0	330	3·0	250	13·0
13·5	15·5	205					2·0	—	3·0	280	2·0	—	13·5
14·0	15·0	200					5·0	—	3·0	—	1·5	85	14·0
14·5	10·0	215					3·0	—	5·5	—	3·0	270	14·5
15·0	10·0	215					0·5	—	5·0	—	8·0	275	15·0
15·5	10·0	205							5·0	—			15·5
16·0	8·5	210							2·5	—			16·0
16·5									4·0	—			16·5
17·0									1·5	—			17·0
17·5									3·5	—			17·5
18·0									3·0	—			18·0
18·5									3·0	—			18·5
	2 theodolites to 4 km. burst		2 theodolites light failed		2 theodolites to 9·5 km. burst		2 theodolites from 1 to 2 km. above 11·5 km. the direction is exceedingly variable. light failed		2 theodolites to 15 km. above 14 km. the direction is exceedingly variable		2 theodolites to 15 km. above 12 km. the direction is exceedingly variable. burst		

Unclassified Ascents.

	Feb. 2, 1907 11.45 a.m.		Feb. 4, 1907 10.44 a.m.		Feb. 14, 1907 2.9 p.m.		Feb. 23, 1907 3.43 p.m.		Mar. 28, 1907 5.55 p.m.		May 20, 1907 4.49 p.m.		
	vel.	dir.	vel.	dir.	vel.	dir.	vel.	dir.	vel.	dir.	vel.	dir.	
Gradient	4·0	45	—	—	8·0	315?	7·0	25	—	—	5·0	20?	Gradient
Surface	2·5	360	4·0	355	3·0	315	3·5	10	2·5	135	4·0	40	Surface
0·5	5·0	50	5·0	350	8·0	325	6·0	20	1·5	70	5·0	40	0·5
1·0	8·5	65	8·0	300	3·5	360	7·0	30	5·5	70	3·0	10	1·0
1·5							[5·0	50]	9·0	80	4·0	10	1·5
2·0									7·5	95			2·0
	cloud		cloud				cloud?		distance		Totland Bay		

	June 17, 1907 7.15 p.m.		Mar. 30, 1908 11.29 a.m.		April 4, 1908 4.14 p.m.		Jan. 19, 1909 10.34 a.m.		Jan. 19, 1909 0.33 p.m.		
	vel.	dir.	vel.	dir.	vel.	dir.	vel.	dir.	vel.	dir.	
Gradient	6·0	270	18·0?	270?	16·0	340	7·0?	315?	7·0?	315?	Gradient
Surface	5·0	250	9·5	235	6·5	305	4·0	330	7·5	325	Surface
0·5	7·5	260	9·0	235	8·5	320	8·5	340	10·5	335	0·5
1·0	6·5	235	11·0	240	13·0	330	10·0	340	8·5	340	1·0
1·5	6·5	235			12·0	340	13·0	305	12·0	330	1·5
2·0					11·0	335	5·0	290	10·0	325	2·0
2·5									9·5	250	2·5
3·0									[15·0	270]	3·0
	cloud		2 theodolites		2 theodolites burst				2 theodolites above 1·1 km.		

The following diagrams represent graphically the relation of wind velocity and direction to height in a certain number of cases. The wind velocity is given in metres per second, and the direction in degrees from the North point through East, South and West; an East wind is therefore represented by 90, a South wind by 180 and so on. Heights are given in kilometres. The short vertical lines on the diagrams show the gradient velocity and direction where this could be calculated. The weather maps show the pressure distribution and the wind at the surface at about the time of each ascent.

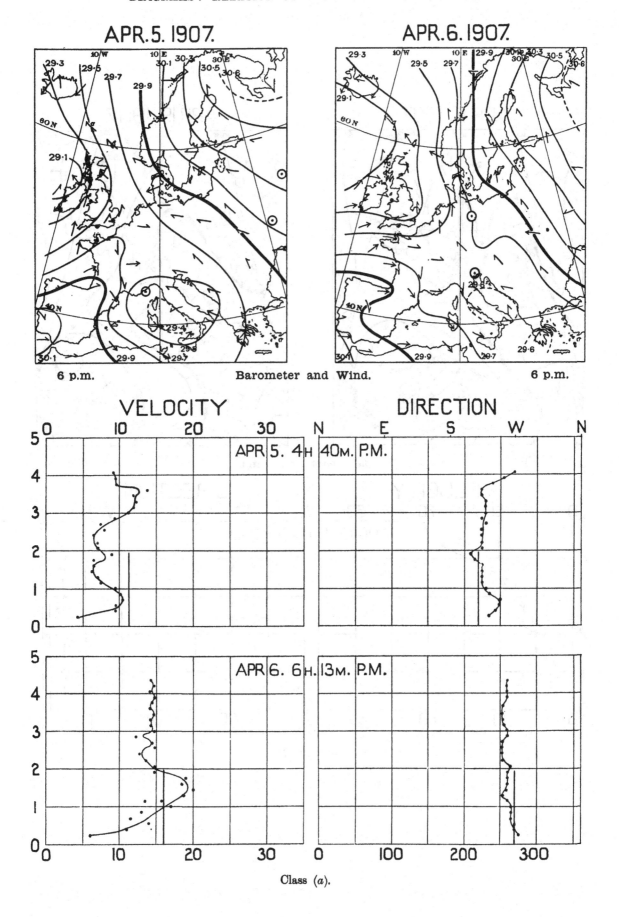

APR. 5. 1907.

APR. 6. 1907.

6 p.m.   Barometer and Wind.   6 p.m.

VELOCITY   DIRECTION

APR 5. 4н 40м. P.M.

APR 6. 6н 13м. P.M.

Class (a).

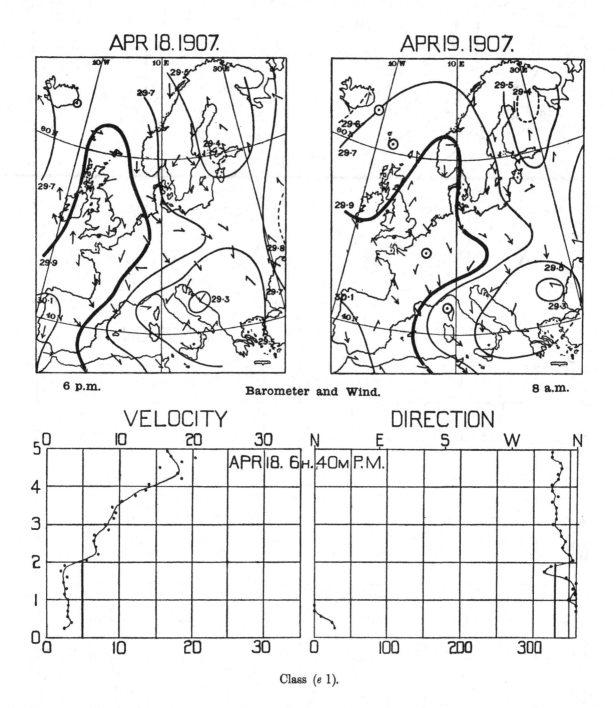

Class (*e* 1).

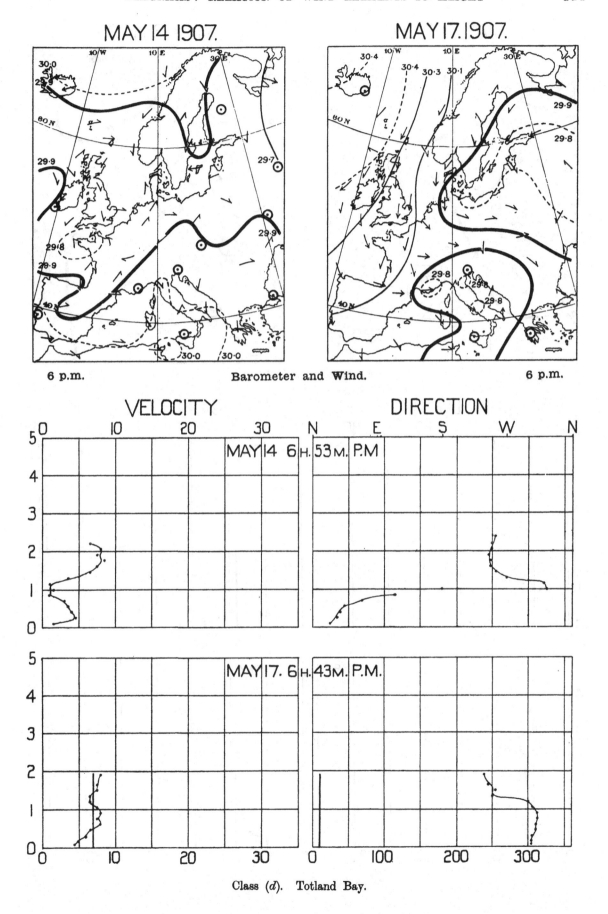

Class (d). Totland Bay.

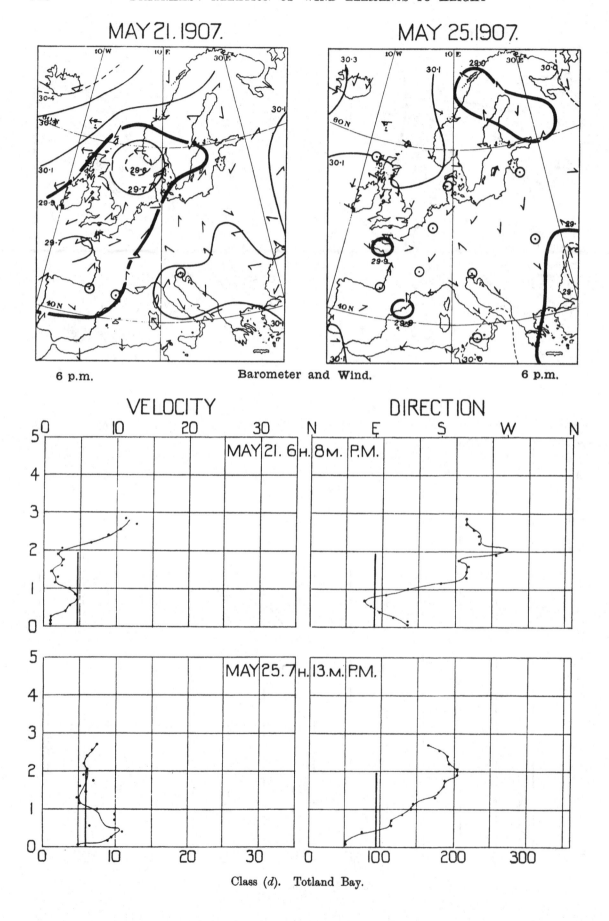

MAY 21. 1907.   MAY 25. 1907.

6 p.m.   Barometer and Wind.   6 p.m.

VELOCITY   DIRECTION

MAY 21. 6 H. 8 M. P.M.

MAY 25. 7 H. 13. M. P.M.

Class (d). Totland Bay.

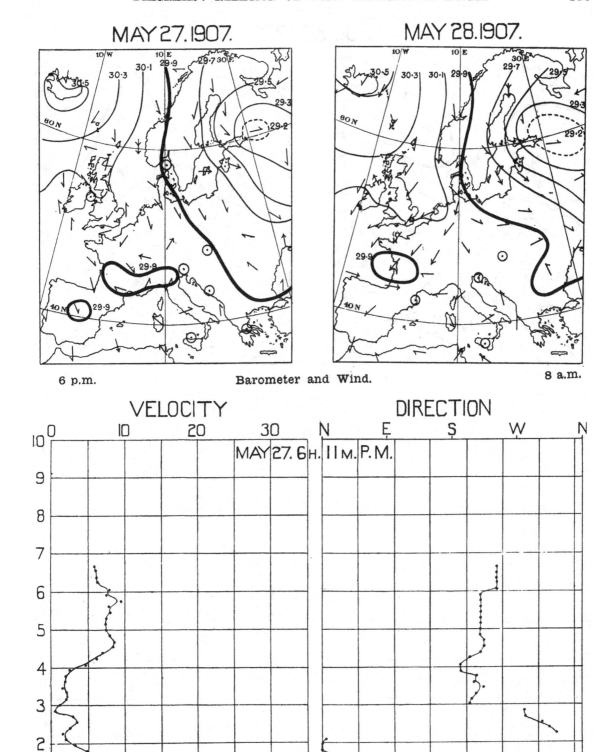

MAY 27. 1907.   MAY 28. 1907.

6 p.m.   Barometer and Wind.   8 a.m.

VELOCITY   DIRECTION

MAY 27. 6H. 11M. P.M.

Class (d).   Totland Bay.

AUG. 28. 1907.

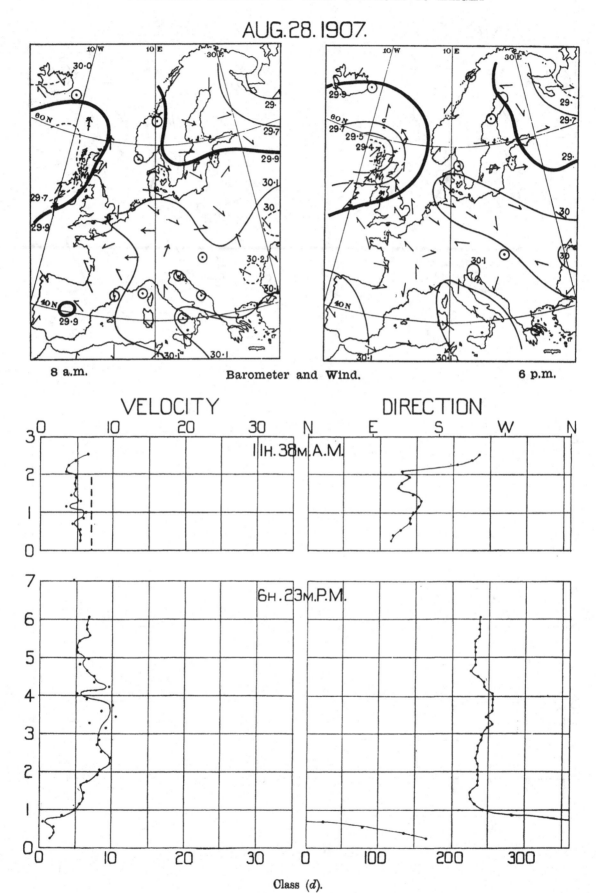

8 a.m.   Barometer and Wind.   6 p.m.

Class (d).

JAN 2. 1908.

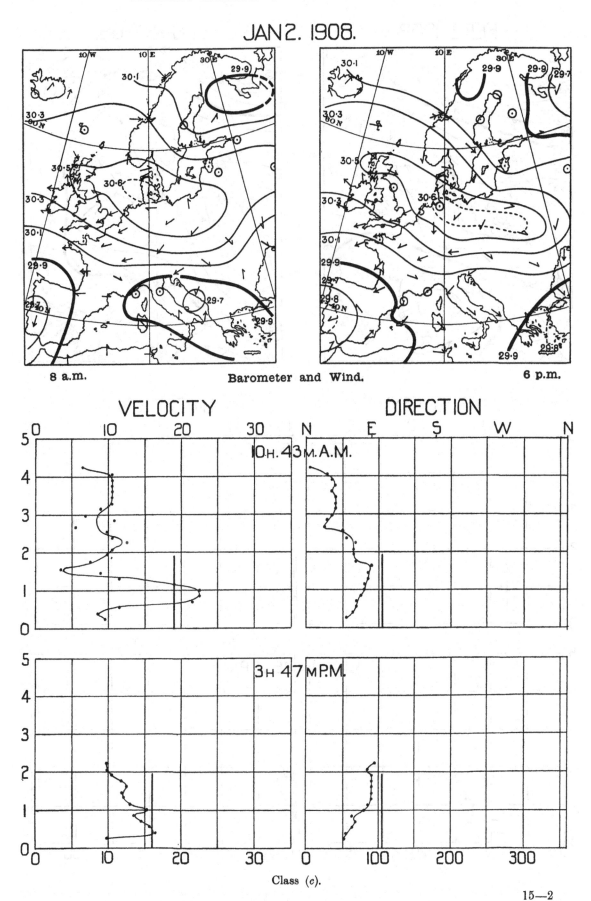

8 a.m.   Barometer and Wind.   6 p.m.

Class (c).

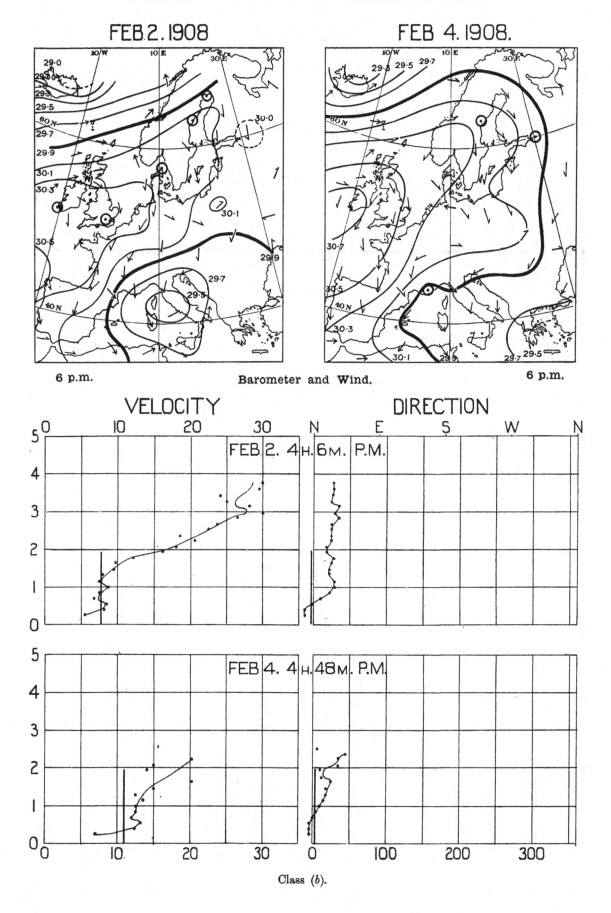

Class (b).

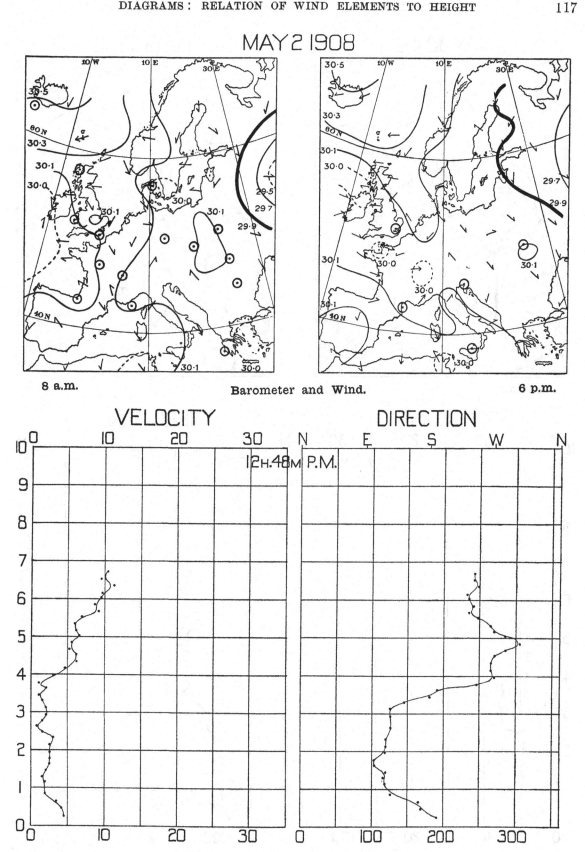

Class (e 2).   Two theodolites to 4 kilometres.

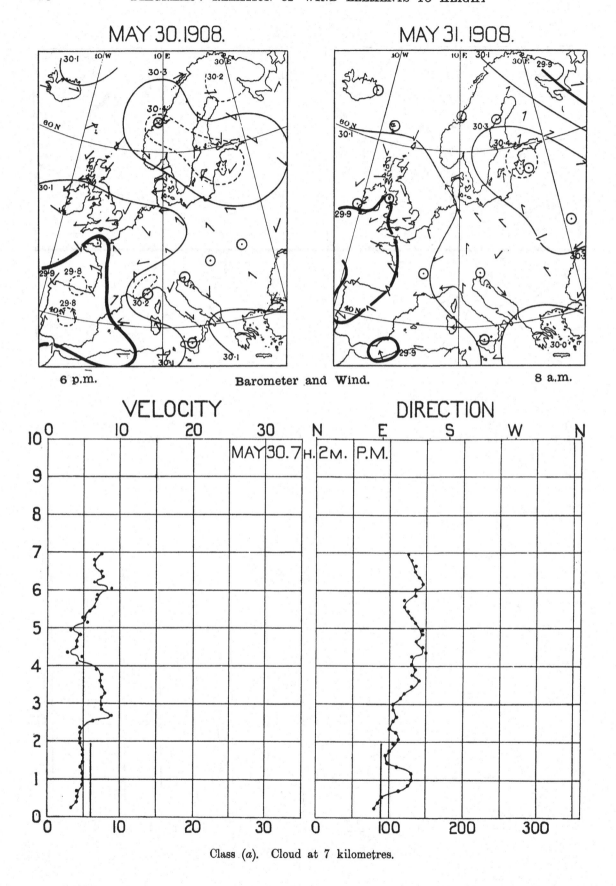

Class (a).   Cloud at 7 kilometres.

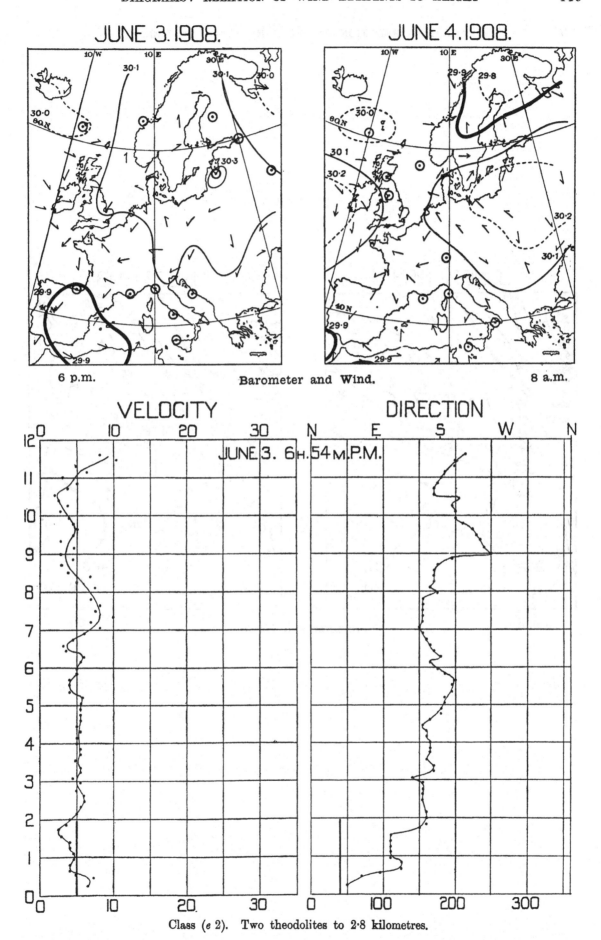

JUNE 3.1908.   JUNE 4.1908.

6 p.m.   Barometer and Wind.   8 a.m.

VELOCITY   DIRECTION

JUNE.3. 6H.54M.P.M.

Class (e 2). Two theodolites to 2·8 kilometres.

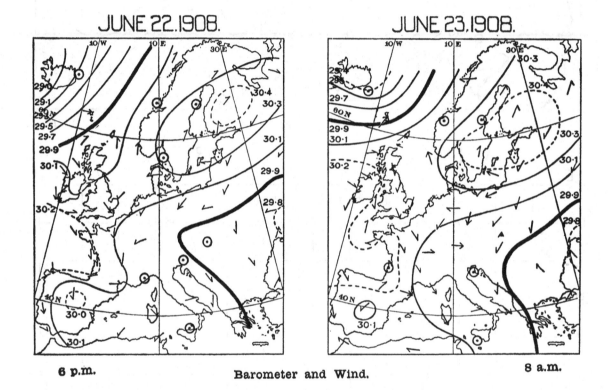

JUNE 22.1908.

JUNE 23.1908.

6 p.m.   Barometer and Wind.   8 a.m.

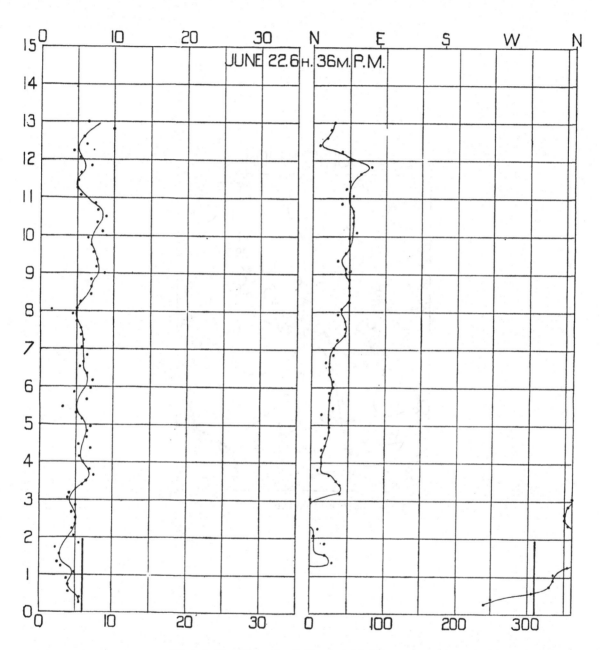

Class (e 1). Two theodolites to 4·3 kilometres.

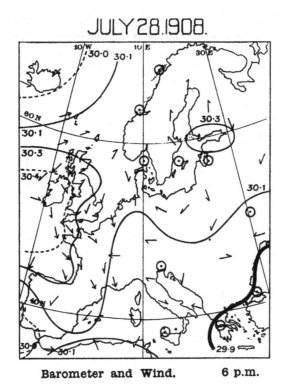

JULY 28.1908.

Barometer and Wind.   6 p.m.

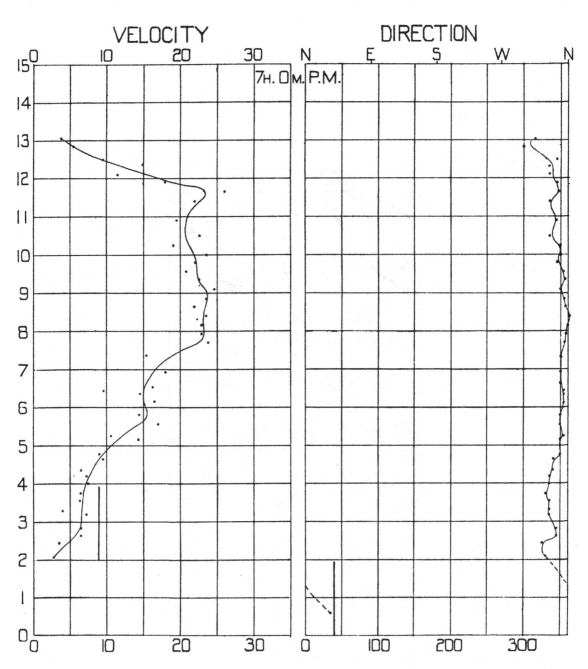

Class (e 1); (f). Two theodolites to 7·7 kilometres.

16—2

JULY 29.1908.

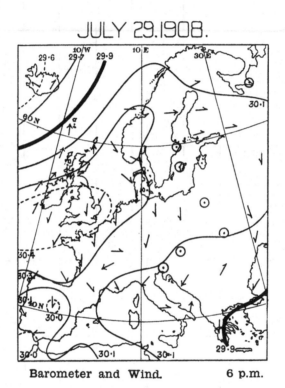

Barometer and Wind.   6 p.m.

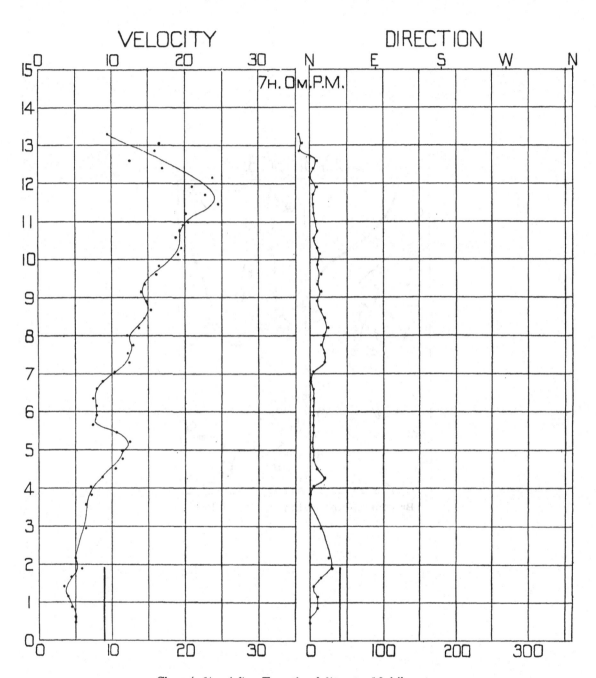

Class (e 1); (f). Two theodolites to 13 kilometres.

JULY 31. 1908.

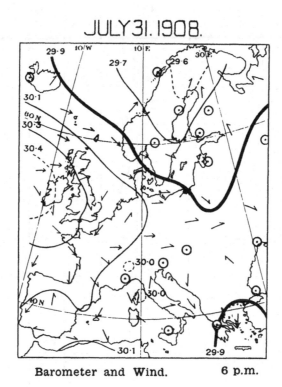

Barometer and Wind.   6 p.m.

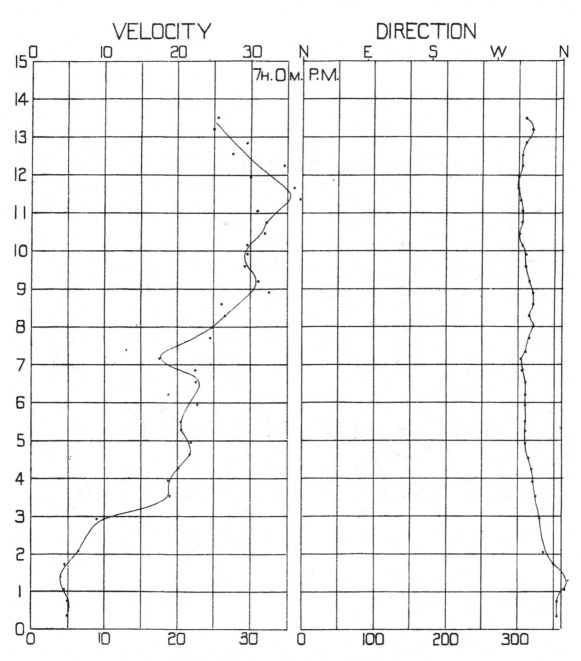

Class (*e* 1); (*f*). Two theodolites to 10 kilometres.

AUG I. 1908.

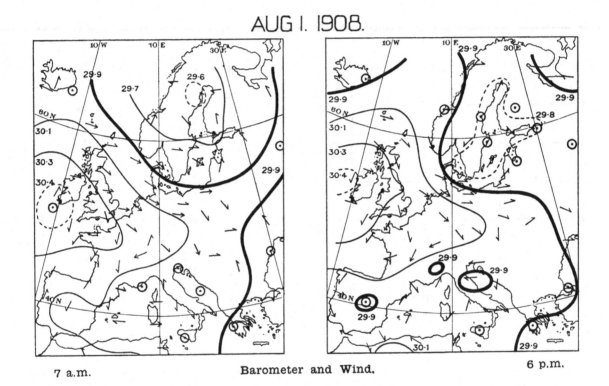

7 a.m.   Barometer and Wind.   6 p.m.

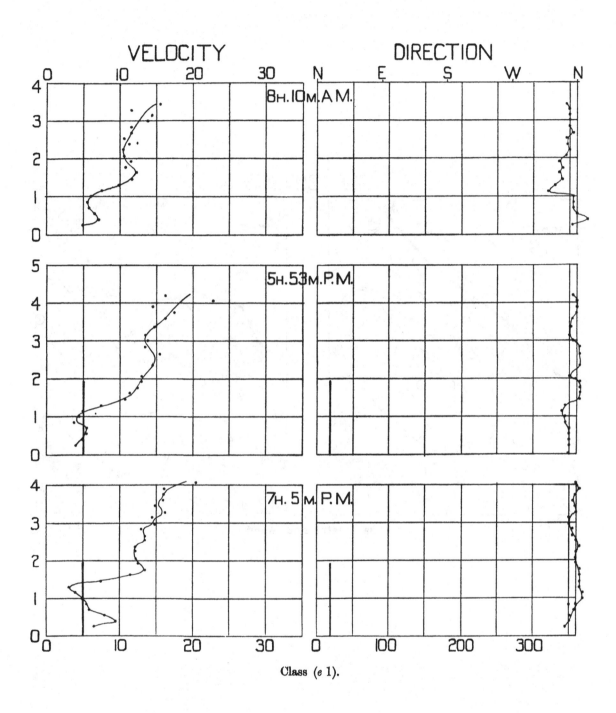

Class (*e* 1).

SEPT. 30. 1908.

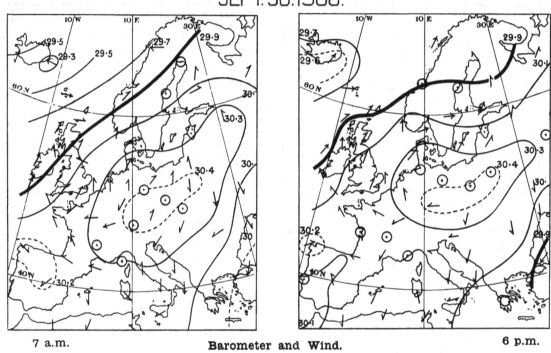

7 a.m.   Barometer and Wind.   6 p.m.

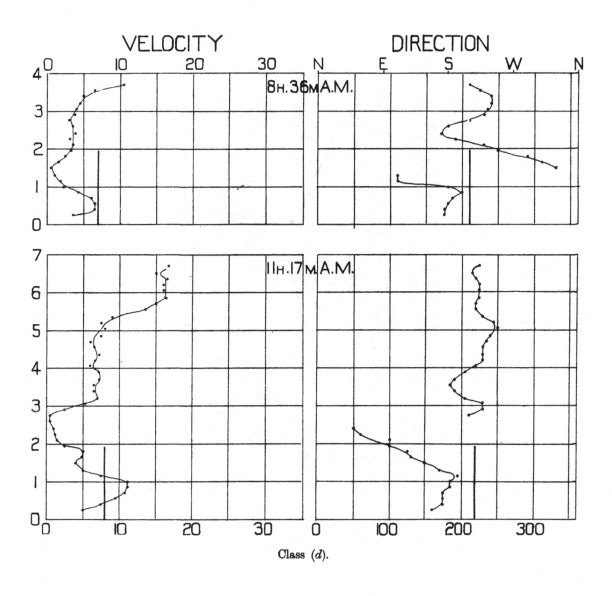

Class (*d*).

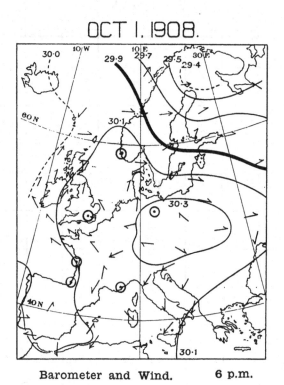

Barometer and Wind.   6 p.m.

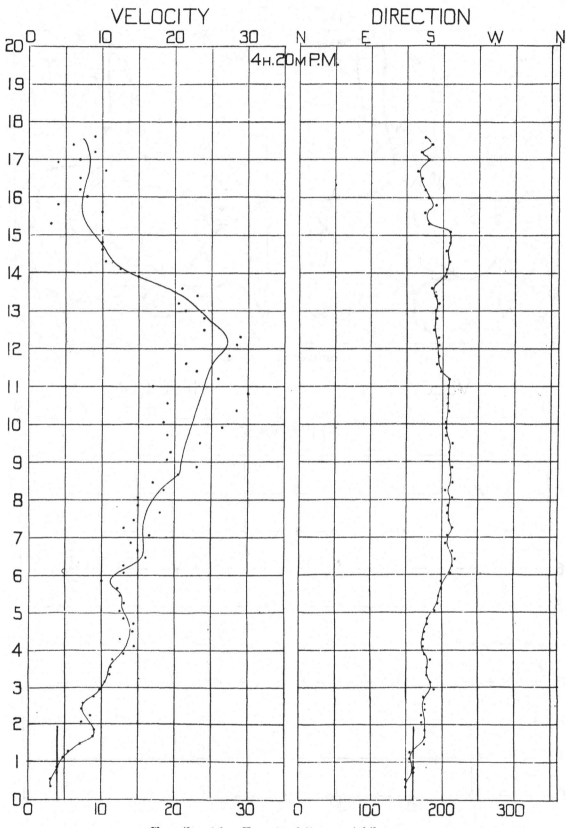

Class (b); (f).   Two theodolites to 4 kilometres.

NOV 6 1908.

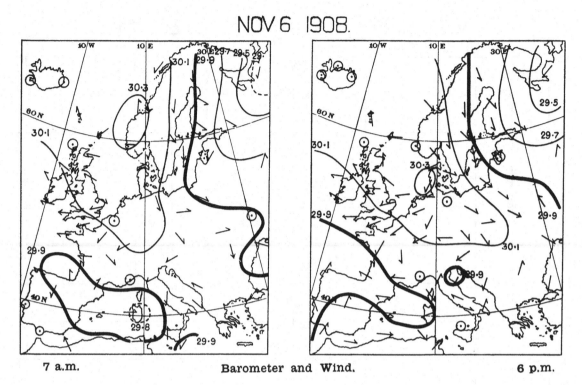

7 a.m.   Barometer and Wind.   6 p.m.

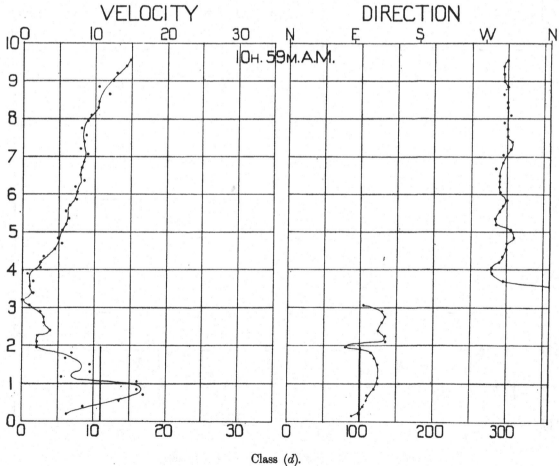

Class (d).

NOV. 7. 1908.

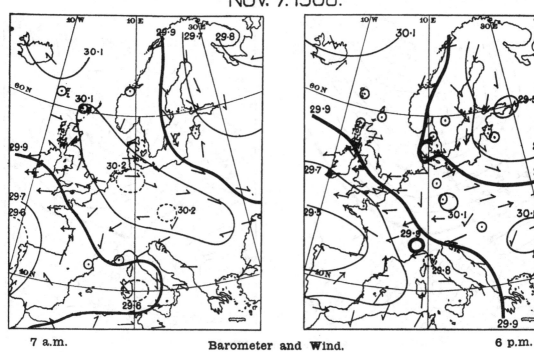

7 a.m.   Barometer and Wind.   6 p.m.

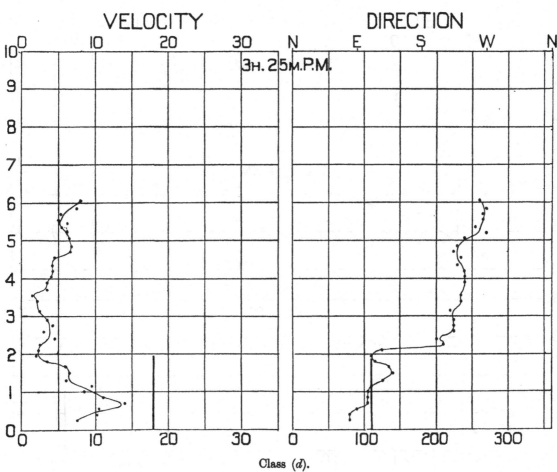

Class (d).

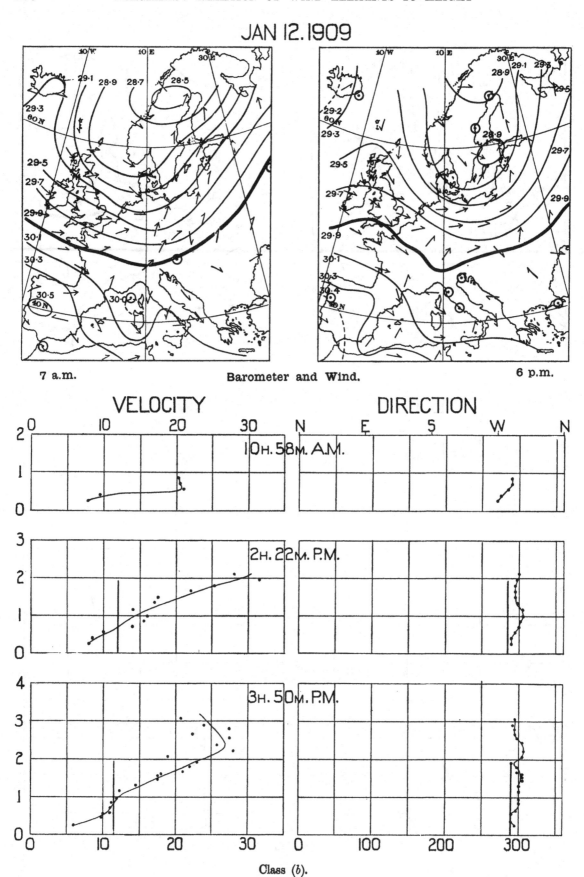

JAN 12.1909

7 a.m.   Barometer and Wind.   6 p.m.

VELOCITY   DIRECTION

10H. 58M. A.M.

2H. 22M. P.M.

3H. 50M. P.M.

Class (b).

FEB 18. 1909.

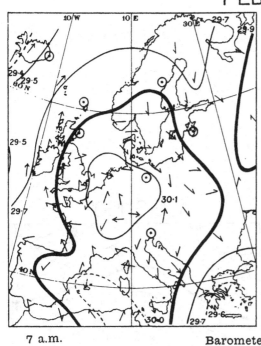
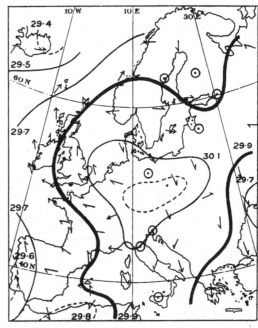

7 a.m.   Barometer and Wind.   6 p.m.

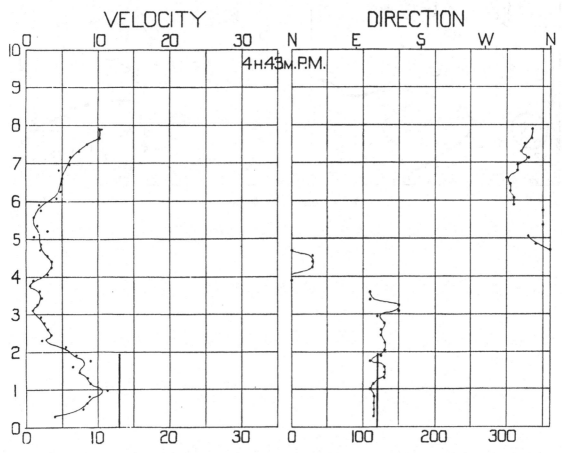

Class (e 1).   Two theodolites to 5·6 kilometres.

c.   18

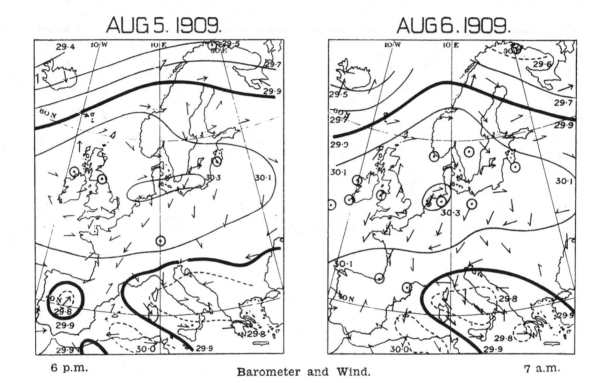

AUG 5. 1909.   AUG 6. 1909.

6 p.m.   Barometer and Wind.   7 a.m.

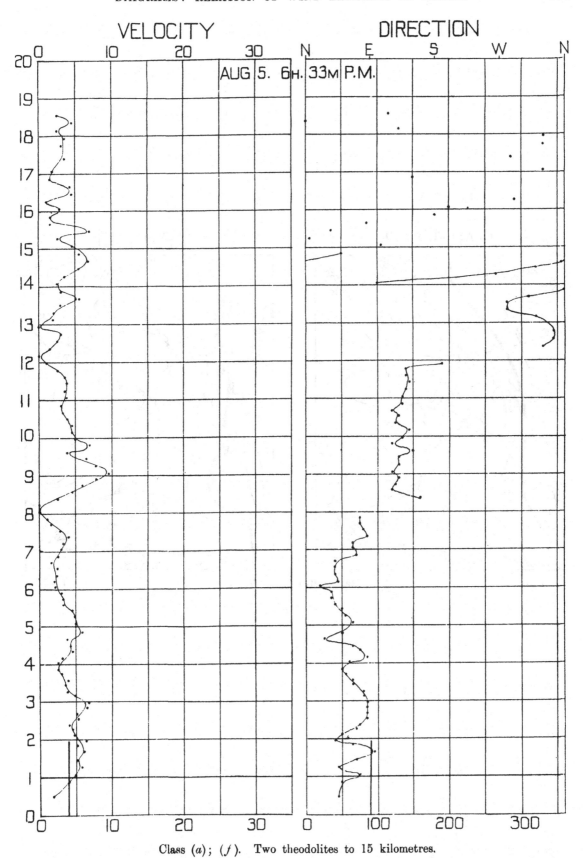

VELOCITY   DIRECTION

AUG 5. 6H. 33M P.M.

Class (a); (f). Two theodolites to 15 kilometres.

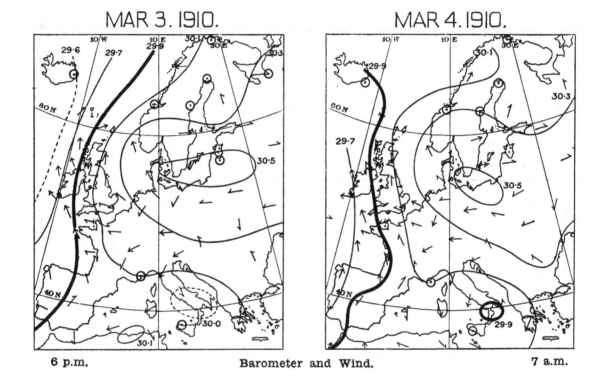

MAR 3. 1910.   MAR 4. 1910.

6 p.m.   Barometer and Wind.   7 a.m.

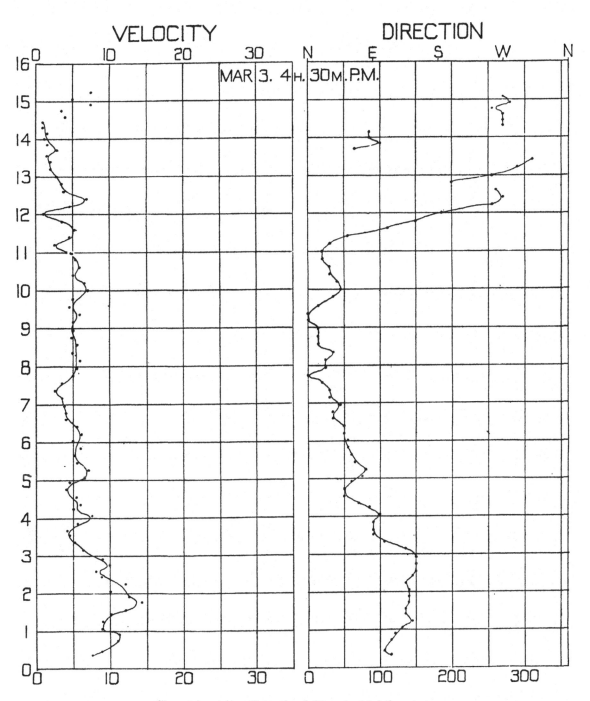

Class (c); (f).   Two theodolites to 15 kilometres.

INDEX

Printed in the United States
By Bookmasters